MATHEMATICAL FOUNDATIONS OF QUANTUM STATISTICS

by

A.Y. KHINCHIN

DOVER PUBLICATIONS, INC.
Mineola, New York

Published in Canada by General Publishing Company, Ltd., 30 Lesmill Road, Don Mills, Toronto, Ontario.
Published in the United Kingdom by Constable and Company, Ltd., 3 The Lanchesters, 162–164 Fulham Palace Road, London W6 9ER.

Bibliographical Note

This Dover edition, first published in 1998, is an unabridged and unaltered republication of the edition published by The Graylock Press, Albany, N.Y., in 1960. It was translated from the first Russian edition published in 1951. The translation was edited by Irwin Shapiro.

Library of Congress Cataloging-in-Publication Data

Khinchin, Aleksandr I͡Akovlevich, 1894–1959.
[Matematicheskie osnovanii͡a kvantovoĭ statistiki. English]
Mathematical foundations of quantum statistics / by A.Y. Khinchin — Dover ed.
p. cm.
Originally published: Albany, N.Y. : The Graylock Press, 1960.
Includes bibliographical references and index.
ISBN 0-486-40025-5
1. Quantum statistics. I. Title
QC174.4.K413 1998
530.13'3'0151—dc21 97–47143
CIP

Manufactured in the United States of America
Dover Publications, Inc., 31 East 2nd Street, Mineola, N. Y. 11501

EDITOR'S PREFACE

I would like to express my gratitude to E. J. Kelly, Jr., M. D. Friedman, W. H. Furry and A. H. Halperin whose translation of the original Russian text made possible this English edition. I should also like to thank the Air Force Cambridge Research Center for partial support of this work.

The translation follows the Russian edition closely. Minor errors in the original text have been corrected without comment, and a few remarks irrelevant to the subject matter have been deleted.

Translations of two articles by Khinchin, which extend his work on statistical mechanics, have been added to the book as Supplements V and VI. However, no attempt has been made to make the terminology and symbolism fully consistent with that of the book proper.

For convenience, all works referred to in the text have been listed together on p. 231.

IRWIN SHAPIRO

CONTENTS

CHAPTER V

Foundations of the Statistics of Material Particles

CHAPTER VI

Thermodynamic Conclusions

PREFACE

In my book [1],* which is devoted to the foundations of *classical* statistical mechanics, I point out (page 7) that the same method may, in principle, be applied to the construction of the mathematical foundations of *quantum* statistics. However, since all aspects of this method must undergo certain changes in form, I decided to write a special monograph devoted to the foundations of quantum statistics. This plan took almost ten years to complete, partly because of the burden of other work, and partly because of the difficulty in applying the method: Inclusion of the "new" statistics (symmetric and antisymmetric) required a more serious modification of the method than I had originally thought necessary.

Despite more or less significant alterations of a technical nature, the central idea of the method remains unchanged. In the area of quantum statistics, I show that a rigorous and systematic mathematical basis of the computational formulas of statistical physics does not require a special unwieldy analytical apparatus (the method of Darwin-Fowler), but may be obtained from an elementary application of the well-developed limit theorems of the theory of probability. Apart from its purely scientific value, which is evident and requires no comment, the possibility of such an application is particularly satisfying to Soviet scientists, since the study of these limit theorems was founded by P. L. Chebyshev and was developed further by other Russian and Soviet mathematicians. The fact that these theorems can form the analytical basis for all the computational formulas of statistical physics once again demonstrates their value for applications.

This monograph, like my first book, is devoted entirely to the mathematical method of the theory and is in no way a complete physical treatise. In fact, no concrete physical problem is considered. The book is directed primarily towards the mathematical reader. However, I hope that the physicist who is concerned with the mathematical apparatus of his science will find something in it to interest him.

29 August 1950 A. KHINCHIN

* Numbers in square brackets refer to the references listed at the end of the book.

INTRODUCTION

§1. The most important characteristics of the mathematical apparatus of quantum statistics

The transition from classical to quantum mechanics involves a basic change in the fundamental ideas and concepts of this science. It is therefore not surprising that the mathematical apparatus of statistical mechanics should undergo a significant change in the transition to the concepts of quantum physics. In most cases this change is expressed in a generalization or a refinement of the mathematics, but sometimes the introduction of essentially new mathematical ideas is required. We begin with the enumeration of those new concepts of quantum physics which have the greatest effect on its *statistical* apparatus.

First, we recall two facts which significantly change the external appearance of the mathematical apparatus of statistical physics yet do not have a profound effect on its content: 1) Some physical quantities have *discrete spectra* (denumerable sets of possible values). This fact has only a superficial effect on the mathematical apparatus. It merely requires that finite sums or infinite series be used in place of the usual integrals of classical mechanics. 2) Physical quantities, in addition to depending on the usual Hamiltonian variables of classical mechanics, depend on "spin" variables which are specific to quantum physics and have no analog in the classical theory. This fact also causes no change in the basic ideas of statistical physics, but only complicates the calculations slightly in certain cases. In order not to obscure the fundamental concepts of the theory with details which are not of primary importance for the mathematical method, we avoid mentioning spin wherever possible in this book.

A new aspect of quantum mechanics, which is not present in classical physics, is the statistical nature of its assertions. In classical mechanics, the state of a system uniquely determines the values of all the physical quantities associated with it. Since every such quantity is a function of the Hamiltonian variables, specifying the values of the latter is equivalent to specifying the state of the system. In quantum mechanics, the state of a system defines the physical quantities only as *random variables*, i.e., it determines the *laws of distribution* obeyed by the physical quantities and not their values. This essentially statistical feature of quantum mechanics is independent of and distinct from the statistical aspects of the special methods of statistical physics. In statistical physics the mean value of a physical quantity is found by averaging the quantity *over different states of the system*. In quantum mechanics, however, we speak of the mean value of a quantity *in a certain definite fixed state*. Therefore, quantum statistics, as distinct from the classical theory, is a statistical theory in a double sense of the word. It is very important to distinguish carefully between the concepts and computational methods of quantum mechanics on the one hand and those of statistical physics on the other. We introduce a special terminology and system of notation for each of them, and we rigorously avoid confusing the two sets of ideas since they effectively have nothing in common, except that both are statistical in nature.

This double statistical character of quantum statistics has a somewhat greater effect on the mathematical apparatus than the two facts mentioned above. However, even here the necessary changes do not affect the basis of the method. The intrinsic statistical nature of quantum mechanics, because it is completely independent of the special methods of statistical physics, does not cause any change in the essence of these methods. Only a new superstructure is required, and this requirement merely changes the appearance of the final result.

Finally, we must consider in detail two new features of quantum mechanics which have a much more profound effect on the apparatus of statistical physics. In some applications these features require a qualitative change in the apparatus.

The first of these features involves the so-called "new" statistical schemes (Bose-Einstein and Fermi-Dirac) which do not and cannot have analogs in classical statistical mechanics. In principle, the situation just alluded to is also possible in the classical theory: It is only a question of the necessity of forming mean values of physical quantities by averaging over some (small) fraction of the number of states of the system which have a given total energy. In the classical theory such a reduction of the averaging manifold becomes necessary whenever the equations of motion have a single-valued integral which is independent of the energy integral (see [1; §10, p. 47]). However, such a necessity rarely arises in practice, since under ordinary conditions integrals of this kind either do not occur at all; or, if they do occur, the averages over the reduced manifold prove to be practically the same as the averages over the original complete manifold.

The transition to the "new" statistics signifies just such a reduction of the manifold of "accessible" states of the system over which the averaging must be carried out. The reduction is necessary because of the existence of a certain single-valued integral of Schrödinger's equation. (This quantum-mechanical equation describes the evolution in time of the state of a system and thus replaces the "equations of motion" of classical mechanics.) The existence of this integral (we call it "the index of symmetry" in the following) is the rule, not the exception. Also, the mean values obtained by averaging over the reduced manifold differ from those obtained by averaging over the original complete manifold to such a degree that it is absolutely necessary to calculate these differences. In classical mechanics the equations of motion cannot have integrals which are in any way analogous to this "index of symmetry". This index describes a specific feature of quantum mechanics.

In constructing a general statistical theory, this necessity to reduce the averaging manifold leads to an essential complication of the mathematical apparatus. The local limit theorems of the theory of probability remain the fundamental basis as before. However, even in the simplest case of systems consisting of particles of a single type, which obey symmetric or antisymmetric statistics, it is convenient to use two-dimensional limit theorems instead of a one-dimensional theorem. The use of a one-dimensional limit theorem only suffices for the case of complete statistics (i.e., for the case of the classical "Maxwell-Boltzmann scheme"). The reduction of the computational problems of statistical physics to those of establishing limit theorems of the theory of probability also undergoes significant changes. In addition, the need to carry out all computations on an extremely general basis, which simultaneously includes all three basic statistical schemes, naturally makes the exposition more complicated.

The second specific feature of quantum mechanics which exerts a substantial influence on the methods of statistical physics involves the problem of the "suitability" of the mean values given by these methods; that is, the question of whether these mean values can be verified by experiment. (This will be the case if the dispersions are small.) To answer this question, it is customary in classical statistical mechanics to formulate so-called ergodic hypotheses or theorems. These state that, on the average, a system, whose evolution in time is governed by the equations of motion, remains in different parts of a given manifold of constant energy for fractions of the total time interval which are proportional to the volumes of these parts. Therefore, if we observe any physical quantity associated with a given system over a definite time interval, the arithmetic average of the results of a sufficiently large number of measurements will, as a rule, be close to the (theoretical) statistical average. It is well-known that in classical statistical mechanics no attempt at such an "ergodic" approach to establishing the suitability of theoretical averages has yet led to any completely satisfactory solution (despite a series of remarkable isolated successes). However, in quantum statistical mechanics such an "ergodic" approach turns out to be impossible in principle. A classical mechanical system changes its state according to the equations of motion and during the course of time its state, at least in principle, can approach as closely as desired to any previously specified state which has the same total energy. This statement is used as a basis for the attempt to compare theoretical averages of physical quantities, taken over all possible states of an isolated system, with data obtained from measurements of the corresponding quantities of the same system made at different times in its evolution. In quantum mechanics the situation is completely different. If a system has a definite (fixed) total energy (i.e., if the system is in a "stationary" state) and evolves according to Schrödinger's equation, then the distribution law of any physical quantity associated with the system remains invariant in time. (We prove this in Chapter II, §5.) But in quantum mechanics the state of a system determines only the distribution laws obeyed by the physical quantities associated with the system. We must therefore suppose that the state of a system, which has a definite total energy, does not in general change with time. Hence, the average of a sequence of measurements performed on such a system (even if a sequence of this kind were possible without radical disruption of the state of the system by each individual measurement) should yield a result which has nothing in common with the theoretical statistical average, since the latter is obtained by averaging the quantity over *all* states which have the same total energy as the given system.

Thus, regardless of how we appraise the effectiveness of ergodic methods in classical statistical mechanics, in quantum statistics they are in principle of no value in establishing the suitability of the theoretical mean values of physical quantities. (See [2].) The "time averages" of such quantities, in virtue of the above discussion, will, as a rule, be quite different from the theoretical mean values. Therefore, in choosing a mathematical apparatus in quantum statistics we must consider the need to find other methods for establishing the suitability of mean values. As we shall see, this requires a very accurate estimate of the remainder terms in the relevant limit theorems of the theory of probability. In particular, the accuracy must be significantly improved compared to that required for estimating mean values.

We wish to emphasize once again that despite the various changes necessary in the mathematical apparatus the central idea of our methods remains unchanged in the transition from classical to quantum physics. This idea consists in the systematic application of the asymptotic formulas of the theory of probability to all the calculations of statistical physics. These formulas represent a general study of mass phenomena, and provide a rigorous mathematical foundation for statistical physics. Therefore, the creation of a special analytical apparatus is unnecessary.

§2. Contents of the book

We mentioned in the Preface that this book is intended for two categories of readers: physicists interested in the mathematical foundations of their science and mathematicians who wish to become acquainted with physical applications of mathematics. As a rule, these two groups approach the reading of a book with different backgrounds. Therefore, to provide both types of readers with the minimum amount of material necessary to master the basic sections of the book, we have expanded the introductory part somewhat. Two long chapters (the first and the second) are devoted entirely to preliminary material and the treatment of the problems of quantum statistics does not begin until Chapter III.

The first chapter contains a discussion and complete proofs of those limit theorems of the theory of probability which are used in the main sections of the book. We refer here to local limit theorems for sums of identically distributed random variables that can assume only non-negative integral values. It is well-known that the general conditions for the applicability of theorems of this type were found only quite recently by B. V. Gnedenko and his students. Chapter I contains complete proofs of the local theorems for the one-dimensional and two-dimensional cases. The fundamental method of Gnedenko is used in these proofs. However, in view of the applications to be made of these theorems, the calculations are carried out in somewhat more detail in order to obtain not only asymptotic formulas, but also accurate estimates of the remainder terms. Thus, this chapter contains a certain element of novelty even for a mathematician whose specialty is the theory of probability. For mathematicians of other specialties, and also for physicists, it will doubtless be completely new. Readers who are not interested in the details of the proofs of the limit theorems should not read the first chapter thoroughly but merely become acquainted with the statements of the theorems which are given at the end of §§4 and 5.

The second chapter introduces the necessary preliminary concepts of quantum mechanics. The educated physicist will, as a rule, find it superfluous. We suggest that he only glance at it to familiarize himself with the terminology and the system of notation used in the remainder of the book. The mathematician will probably find it necessary to read this chapter. However, we must caution him that familiarization with its contents cannot replace a preparatory mastery of the fundamental ideas of quantum physics which can even be obtained from literature of a more or less popular character. Chapter II can not be considered either as a short course or as a synopsis of a course in quantum mechanics. The choice of material is not intended to be exhaustive, but is determined solely by an interest in the special problems discussed in the remaining chapters. In particular, the second chapter is concerned almost exclusively with the mathematical apparatus of quantum mechanics. The physical content of the subject has not been emphasized. From a formal point of view this chapter contains everything necessary to understand the following sections of the book. However, for the reader who is totally unacquainted with the ideas of quantum physics, it will be of little value: A knowledge of its purely formal content will only leave the reader in mid-air. (It is sufficient to point out that in the entire chapter not a single experiment is mentioned.) We repeat, therefore, that the mathematician approaching the study of our book must have at least a modest acquaintance with the general ideas of quantum physics. As we have said, this acquaintance may be obtained from very elementary sources. If the principal physical ideas of quantum physics are already known to the reader, then our second chapter will easily raise this knowledge to the mathematical level necessary for an understanding of the following chapters.

The third chapter contains an exposition of the general ideas and the basis of the computational methods of quantum statistics. In classical mechanics the statistical theory is used primarily to investigate the statistics of various physical quantities associated with a system of given total energy. Similarly, in quantum statistics the distribution laws of various physical quantities are studied for a system that has a *definite* total energy. Thus, at least at the outset, only those states of a system will be considered in which the total energy has a definite value. These states are described by eigenfunctions of the total energy operator $\mathfrak{H}$.

In classical mechanics the set of states in which the total energy of the system has a given constant value forms some "surface of constant energy" in the phase space. In quantum mechanics the analogous set of states is the linear manifold $\mathfrak{M}$, whose elements are eigenfunctions of the operator $\mathfrak{H}$ belonging to some definite eigenvalue of this operator. For systems considered in statistical physics this manifold always has a finite but very large dimension [a high degree of degeneracy (multiplicity) of the eigenvalues]. In classical mechanics mean values of physical quantities are obtained by averaging over a given surface of constant energy (or parts of it, if in addition to the energy integral there are other single-valued integrals of the motion). Similarly, in quantum statistics the averaging is performed over the manifold $\mathfrak{M}$ or over parts of it. In fact, as we remarked in §1, it is necessary to make a significant reduction in this manifold for the majority of systems considered in statistical physics. In these cases only symmetric or antisymmetric eigenfunctions are admissible. Thus, in the statistical problems of quantum physics it is necessary to develop computational methods for three fundamental statistical schemes: complete, symmetric and antisymmetric. For this purpose, we establish first a particular complete orthogonal system of eigenfunctions for each of these three schemes. These functions have great importance for all that follows, and we call them the fundamental eigenfunctions. The states which are described by these eigenfunctions are called the fundamental states of the system.

Further, we introduce the notion of "occupation numbers" which is of basic importance in quantum statistics. Each of these specifies the number of particles of the system which is found in a particular state. The fundamental states chosen are especially convenient for statistical calculations because in each of these states the occupation numbers have definite (fixed) values. Thus, some definite set of occupation numbers corresponds to each fundamental state in any of the three statistical schemes. Conversely, one or several fundamental states correspond to each set of the occupation numbers. The number of fundamental states corresponding to a given set of occupation numbers is different for the three basic statistical schemes. This difference is the most important consequence of the statistical dissimilarity of these schemes.

Many of the most important physical quantities studied in statistical physics have a "sum" character, i.e., they are sums of quantities each depending on the state of only one of the particles which compose the system. The mean value of a "sum function" can be written down immediately from a knowledge of the mean values of the occupation numbers. If in addition to the mean values of the occupation numbers we are able to find the mean values of their pairwise products, then we can immediately write the dispersion of an arbitrary sum function. These facts explain why authors of systematic expositions of quantum statistics consider the determination of the mean values of the occupation numbers to be their most important initial task. It should be noted, however, that the mean values of the occupation numbers determine directly the mean values only of sum functions. Even though the sum functions are the most important functions they do not exhaust all quantities which can be of interest in statistical physics. Any quantity which depends symmetrically on the states of the particles which compose the system can be of interest in statistical physics. While sum functions are the simplest and most frequently encountered of these symmetric functions, they evidently do not exhaust the set. (Thus the dispersion of a sum function is symmetric but is obviously not a sum function.) From the mathematical point of view it would no doubt be an interesting and worthwhile task to consider a broader class of problems. However, we must note that the limit laws for symmetric functions of a large number of random variables are still completely undeveloped. [EDITOR'S NOTE: In several later articles Khinchin did consider a broader class of symmetric functions. This work has been included here as Supplements V and VI.]

At the end of the third chapter we show that the problem of establishing the suitability of microcanonical averages can be reduced to that of estimating the microcanonical dispersions of the corresponding physical quantities. In particular, we derive an expression for the dispersion of sum functions which is valid for all three statistical schemes.

After establishing the foundations of the statistical methods of quantum physics, we give a concrete structure to quantum statistics in the fourth and fifth chapters. The fourth chapter is devoted to the statistics of photons, and the fifth to the statistics of material particles (i.e., particles with non-zero "rest mass"). We start with photons solely for pedagogical reasons. It is well-known that the number of photons constituting a given system is not constant, but can change with time. This makes the statistics of a "photon gas" substantially simpler than the statistics of systems consisting of material particles. Therefore, we develop all the computational methods first using this simplest example for which a one-dimensional limit theorem suffices. We hope that the reader masters this chapter before passing to the more complicated case of material particles. He will then be acquainted with the fundamental ideas of the method and the purely technical complications encountered in Chapter V will not cause him any great difficulty.

The derivation of the fundamental computational formulas is carried out in completely parallel fashion in these two chapters. The dimension of the linear manifold of eigenfunctions of the operator $\mathcal{H}$ which belong to a given eigenvalue E is a function of E, and is called the structure function of the system. (In the case of material particles the structure function also depends on the number of particles composing the system.) The first step of the derivation is to determine the exact expressions for the mean values of the occupation numbers and their pairwise products in terms of the structure function. These expressions are very simple but are different for the different statistical schemes. They enable us to reduce completely the problem of finding asymptotic estimates of the mean values of the occupation numbers and their pairwise products to that of finding approximate expressions for the structure function. The second step is to express the structure function for each case in terms of the distribution law of a random variable which is defined as the sum of a very large number of mutually independent and identically distributed random variables. In general, these distribution laws are multi-dimensional. Only in the problem of photons are they one-dimensional. Finally, in the third and last step of the computation the limit theorems derived in Chapter I are applied to obtain asymptotic expressions for these distribution laws. This gives us convenient, and at the same time very accurate, approximate expressions for the structure function and, hence, for the mean values of the occupation numbers and their pairwise products. Using these expressions and the methods developed in Chapter III, we easily find approximate expressions for the mean values and the dispersions of sum functions which are just as accurate. The precision obtained in this case turns out to be perfectly adequate to establish the suitability of microcanonical averages of physical quantities which are sum functions.

In the sixth and last chapter the results found previously are used to define the concept of entropy and to establish by statistical methods the basis of the second law of thermodynamics and, hence, of all of thermodynamics.

In the "supplements" which complete the book, we consider several topics of significant interest that are somewhat peripheral to the main line of development of the theory. Therefore, we prefer to assign to them a special place at the end of the book.

We must still add some general remarks concerning the method of exposition adopted in this book. These remarks are necessary for a broader understanding of the material.

1. In the best expositions of statistical physics following the first works of Darwin and Fowler, basic use has been made of the *mean* values of physical quantities rather than of their *most probable* values. Prior to this period, the most probable values were invariably used. In regard to this change, we will mention two very basic facts: i) In those cases where the mean and most probable values of quantities are considerably different from each other, the mean values always play the deciding role in descriptions of macroscopic phenomena. ii) The methods adopted in statistical physics for the calculation of the most probable values always deserve the valid criticism of being mathematically incomplete. On the other hand, the method of calculation of mean values, formulated by Darwin and Fowler, is faultlessly rigorous in mathematical respects.

In the following, we always speak of the *mean* values of physical quantities. It is true that for most quantities of interest in statistical physics, the differences between the mean and most probable values are sufficiently small so that in practice they may be neglected. However, rather than demonstrate rigorously that these differences are negligible (a conclusion considered obvious by those who use most probable values) we present a sufficiently well-developed theory of mean values so that a consideration of most probable values is unnecessary.

The sole advantage of the (mathematically unrigorous) calculations of most probable values consists in their incontestable relative simplicity. There is no doubt that the method of Darwin and Fowler, which is based on a specially constructed analytical apparatus, is very complicated mathematically. This explains its relative unpopularity among physicists. But, as we mentioned in the Preface, the primary purpose of this book is to show that there is no need for a special analytical apparatus to justify rigorously the methods of calculation of mean values of physical quantities. The calculation leads to a completely elementary application of general and well-known limit theorems of the theory of probability. Thus, the last purely practical objection to the transition from most probable values to mean values can be eliminated.

2. The reader who is acquainted with this subject will probably notice that in our book, in contrast to the majority of contemporary expositions of this subject, no mention is ever made of the so-called "statistical ensembles". The systematic use of this term usually indicates that the physical system being studied is considered as an element of some set of physical systems which have the same structure as the given system but are in different states. However, it should be understood that in all statistical theories, a physical quantity is given a value which is obtained by averaging the quantity over all different states of the observed system. Such a value, in general, corresponds to the mean value of the results of a large number of experimental observations of the quantity. This connection of the statistical theory with the physical world is discussed in great detail and is often emphasized throughout our book. It appears to us, however, that the systematic assignment of a physical system of the same structure to each of the possible states of the given system (thus forming the "ensemble" introduced by many authors) is completely superfluous and only hinders an understanding of the theory. We prefer to consider the set of states which are possible states for the system (a phase space in classical physics) and not to consider a set of systems which are assigned to these states. The latter approach only complicates the picture of the phenomenon. This same point of view was introduced in our book on the mathematical foundations of classical statistical mechanics. Instead of assigning a whole "ensemble" of systems of the same type to the phase space and then following the evolution of this "ensemble", we simply spoke of the "natural motion" of the phase space itself. (This can be thought of as a space which is continuously being transformed onto itself, the motion being like that of a simple hydrodynamic model.) This description is simple and convincing from both the mathematical and the physical points of view. Only the unnecessary assignment of some physical system to each point of the phase space is lost.

3. In keeping with our pedagogical aim, we choose only the simplest examples and consciously refrain from considering more complicated situations so that the reader can concentrate all of his attention on the mathematical method. In the text proper we limit ourselves to homogeneous systems (i.e., systems consisting of particles of the same structure) and only in Supplement I do we show how our method can be applied to heterogeneous systems. The particles composing each system are assumed to be enclosed in a vessel of constant volume. It is well-known that this leads to a discrete energy spectrum

$$\varepsilon_1 \leq \varepsilon_2 \leq \cdots \leq \varepsilon_r \cdots$$

for the particles. As is usual in such investigations, we also assume that the energy levels ε_r of the particles are integers. In practice, all the energy levels can be made to approximate integers as closely as desired by choosing a sufficiently small unit of energy. Further, as is also customary, we assume that the energy of the system is equal to the sum of the energies of its constituent particles. This means that we neglect the interaction energy of the particles; i.e., strictly speaking, we are limiting ourselves to multiatom ideal gases. It is, of course, impossible to follow this point of view to its logical conclusion, since in the absence of an interaction between them, the particles cannot exchange energy, and the whole statistical problem becomes meaningless. Usually, as a result of this difficulty, it is assumed (in many cases with good reason) that the interaction between the particles is sufficient to guarantee a free exchange of energy between them, but at the same time is weak enough so that for all practical energy calculations it is possible to equate the energy of the system to the sum of the energies of its constituent particles. Thus, the "mixed" terms which express the interaction energy of the particles can be neglected.

The primary purpose of our method is to deduce all the asymptotic formulas necessary for quantum statistics. As usual, these formulas are derived on the assumption that the number of particles N of the given system, its total energy E, and its volume V are infinitely large quantities which maintain constant ratios (a constant mean energy per particle and a constant mean density of the gas). In essence, this means that units of energy and volume are chosen so that the ratios E/N and V/N are neither too small nor too large. All quantities characterizing the given system which depend only on these ratios must therefore be regarded as constants in our asymptotic formulas.

Chapter I

PRELIMINARY CONCEPTS OF THE THEORY OF PROBABILITY

§1. Integral-valued random variables

This book is concerned with the rigorous and detailed mathematical bases of the most important formulas of quantum statistics. These are established with the help of the limit theorems of the theory of probability, since the question of limit theorems of some particular type arises in all cases. For a long time these theorems have been of interest to specialists and in recent years they have been developed significantly, particularly by mathematicians in the U. S. S. R. Nevertheless they are not, as a rule, discussed in textbooks and consequently are little known to a wide circle of scholars. (As an exception we may mention the book by von Mises [3].) Hence, in the present chapter we give both detailed formulations and complete proofs of the limit theorems which are necessary for our development. We assume only that the reader is acquainted with a general text such as Feller [4].

The type of limit theorem we need is distinguished by the following important specific characteristics:

1) We always consider random variables all of whose possible values are integers;

2) All the limit theorems of interest to us are of the *local* type, i.e., we always consider an asymptotic estimate of the probability that the sum of the random variables being studied assume some definite value;

3) We can limit ourselves to sums of mutually independent and identically distributed random variables that have finite moments up to the fifth order inclusive;

4) In all cases we must find not only an asymptotic formula, but also an accurate estimate of the error;

5) Finally, in addition to the one-dimensional limit theorems, we will be equally interested in multi-dimensional limit theorems of the same type (in particular, two-dimensional limit theorems).

Local limit theorems for integral-valued random variables attracted the attention of investigators a relatively long time ago, although much less effort was devoted to them than to theorems of the "integral" type. Thus, in the book by von Mises, one may find rather deep theorems of this latter type. However, a sufficiently general formulation of the problems of interest to us has been achieved only recently. In particular, the limit theorems of the type we require were first proved by B. V. Gnedenko [5] and his students [6]. Although they considered multi-dimensional problems, the fundamental direction of their investigations differs significantly in one way from that which we need: While Gnedenko and his students always sought more general conditions under which the fundamental limiting relationship is valid, we, as stated above, can confine ourselves to a very narrow class of initial distributions. On the other hand, we cannot be satisfied with deriving limiting relationships, but must estimate the resulting error, sometimes rather accurately. Thus, although Gnedenko's methods are completely adequate for our purpose, we need formulations of limit theorems which are somewhat different from those given by him and his co-workers. This is a further reason for our including a chapter containing detailed proofs of the limit theorems we require.

The random variables we must consider in this book always have only *integers* as their possible values. We call such random variables *integral-valued*. Evidently, the distribution law of the integral-valued random variable ξ is completely determined by giving for each integer n the probability

$$\mathbf{P}(\xi = n) = p_n \qquad [p_n \geq 0\ (-\infty < n < \infty);\ \textstyle\sum_{n=-\infty}^{\infty} p_n = 1],$$

that the variable ξ take on the value n. In the future we shall say briefly that the variable ξ *obeys (is subject to) the law p_n or is distributed according to the law p_n.*

If the series

$$\textstyle\sum_{n=-\infty}^{\infty} np_n$$

converges absolutely, its sum is called the mathematical expectation $\mathbf{E}\xi$ of the variable ξ. (Sometimes, instead of mathematical expectation, the term "mean value" of the random variable ξ is used. We carefully avoid this terminology, since the term "mean value" has a completely different meaning in this book.) In general, given an absolutely convergent series

$$\textstyle\sum_{n=-\infty}^{\infty} f(n)p_n ,$$

where $f(n)$ is an arbitrary real or complex function of the integral argument n, we call the sum of the series the mathematical expectation $\mathbf{E}f(\xi)$ of the random variable $f(\xi)$. In particular, the mathematical expectation $\mathbf{E}\xi^k$ of the variable ξ^k (if it exists) is called the *moment of order k (kth moment)* of the variable ξ. The mathematical expectation $\mathbf{E}(\xi - \mathbf{E}\xi)^k$ of the variable $(\xi - \mathbf{E}\xi)^k$ (if it exists) is called the *central moment* of order k of the variable ξ. The central moment of second order

$$\mathbf{D}\xi = \mathbf{E}(\xi - \mathbf{E}\xi)^2 = \textstyle\sum_{n=-\infty}^{\infty} (n - \mathbf{E}\xi)^2 p_n$$

(if it exists) is called the *dispersion (variance)* of the variable ξ and is, along with the mathematical expectation $\mathbf{E}\xi$, one of the most important characteristics of this variable. All the random variables we shall consider actually possess moments of arbitrary order $k \geq 0$. However, we shall see that for the proof of the relevant limit theorems, it is sufficient to assume the existence of moments of only relatively low orders.

If ξ' and ξ'' are integral-valued random variables obeying, respectively, the laws p_n' and p_n'', then the sum $\xi' + \xi'' = \xi$ is an integral-valued random variable. The distribution law p_n of this sum, in addition to depending on the laws p_n' and p_n'', also depends on the form of the mutual dependence of the variables ξ' and ξ''. In particular, if these two variables are mutually independent, then the numbers p_n are very simply expressed in terms of the numbers p_n' and p_n''. Indeed, in order that $\xi = n$, it is necessary and sufficient that $\xi' = k$, $\xi'' = n - k$, where k is any integer. Therefore,

$$\mathbf{P}(\xi = n) = \sum_{k=-\infty}^{\infty} \mathbf{P}(\xi' = k, \xi'' = n - k),$$

and in virtue of the mutual independence of the variables ξ' and ξ'',

$$p_n = \sum_{k=-\infty}^{\infty} p_k' p_{n-k}''.$$

The last equation may be rewritten as

$$p_n = \sum_{k+l=n} p_k' p_l''.$$

In the same fashion if we have s mutually independent integral-valued random variables

$$\xi^{(1)}, \xi^{(2)}, \cdots, \xi^{(s)}$$

with the corresponding distribution laws

$$p_n^{(1)}, p_n^{(2)}, \cdots, p_n^{(s)},$$

then the distribution law p_n of the sum

$$\xi = \sum_{i=1}^{s} \xi^{(i)}$$

may be expressed as

$$(1) \quad p_n = \sum_{k_1=-\infty}^{\infty} \sum_{k_2=-\infty}^{\infty} \cdots \sum_{k_{s-1}=-\infty}^{\infty} p_{k_1}^{(1)} p_{k_2}^{(2)} \cdots p_{k_{s-1}}^{(s-1)} p_\alpha^{(s)},$$

where $\alpha = n - \sum_{i=1}^{s-1} k_i$; or, equivalently,

$$(2) \qquad p_n = \sum_{\beta=n} p_{k_1}^{(1)} p_{k_2}^{(2)} \cdots p_{k_s}^{(s)},$$

where $\beta = \sum_{i=1}^{s} k_i$. The expression for the distribution of the sum of mutually independent random variables in terms of the distributions of the summands is called the *rule of composition* of these distributions. Thus, formulas (1) and (2) express the rule of composition of the distributions of an arbitrary number of integral-valued random variables.

It is known from the elementary theory of probability that the mathematical expectation of a sum of an arbitrary number of random variables is always equal to the sum of their mathematical expectations. If the summands are mutually independent, the analogous law holds true for the products. Finally, the dispersion of the sum is equal to the sum of the dispersions of the summands when the summands are *pairwise* mutually independent.

We must repeatedly consider cases in which the basic element is not one random variable, but a family of several (two, three or more) *mutually dependent* integral-valued random variables $\xi, \eta, \cdots$. For simplicity of notation we consider the case of a pair (ξ, η) of such variables. (All that is said for this case holds true with the corresponding obvious changes for any larger family of variables.) A pair of this type is sometimes called a (two-dimensional) *random vector*. The probability $\mathbf{P}(\xi = l, \eta = m)$ of the simultaneous realization of the equations $\xi = l$ and $\eta = m$ is denoted by p_{lm}. The set of numbers p_{lm} $(-\infty < l, m < \infty)$ forms the *distribution law* of the random pair (ξ, η). If p_n and q_m, respectively, denote the distribution laws of the variables ξ and η, then evidently

$$p_l = \sum_{m=-\infty}^{\infty} p_{lm}, \qquad q_m = \sum_{l=-\infty}^{\infty} p_{lm}. \tag{3}$$

Hence,

$$\mathbf{E}\xi = \sum_{l=-\infty}^{\infty} l p_l = \sum_{l=-\infty}^{\infty} l \sum_{m=-\infty}^{\infty} p_{lm},$$

and analogously,

$$\mathbf{E}\eta = \sum_{m=-\infty}^{\infty} m \sum_{l=-\infty}^{\infty} p_{lm}.$$

It is assumed that all these series converge absolutely.

If $f(\xi, \eta)$ is an arbitrary real or complex function of the variables ξ and η, then the quantity

$$\mathbf{E}f(\xi, \eta) = \sum_{l=-\infty}^{\infty} \sum_{m=-\infty}^{\infty} f(l, m) p_{lm} \tag{4}$$

is called its mathematical expectation if the double series converges absolutely. In particular, from formulas (3) or (4) we obtain expressions for the dispersion of the variables ξ and η in terms of the numbers p_{lm}:

$$\mathbf{D}\xi = \mathbf{E}(\xi - \mathbf{E}\xi)^2 = \sum_{l=-\infty}^{\infty} \sum_{m=-\infty}^{\infty} (l - \mathbf{E}\xi)^2 p_{lm},$$
$$\mathbf{D}\eta = \mathbf{E}(\eta - \mathbf{E}\eta)^2 = \sum_{l=-\infty}^{\infty} \sum_{m=-\infty}^{\infty} (m - \mathbf{E}\eta)^2 p_{lm}.$$

The ratio

$$\mathbf{R}(\xi, \eta) = \mathbf{E}\{(\xi - \mathbf{E}\xi)(\eta - \mathbf{E}\eta)\}/(\mathbf{D}\xi\, \mathbf{D}\eta)^{\frac{1}{2}}$$
$$= [\mathbf{E}(\xi\eta) - \mathbf{E}\xi\mathbf{E}\eta]/(\mathbf{D}\xi\, \mathbf{D}\eta)^{\frac{1}{2}}$$

is called the *correlation coefficient* of the variables ξ and η. The numerator of this ratio can be written in the form

$$\mathbf{E}\{(\xi - \mathbf{E}\xi)(\eta - \mathbf{E}\eta)\} = \sum_{l=-\infty}^{\infty} \sum_{m=-\infty}^{\infty} (l - \mathbf{E}\xi)(m - \mathbf{E}\eta)p_{lm} .$$

From the random pairs (ξ', η') and (ξ'', η'') we may form the pair $(\xi' + \xi'', \eta' + \eta'')$ which is called the sum of the given pairs. In the same fashion, the sum of an arbitrary number of pairs can be defined. Let p_{lm}', p_{lm}'', p_{lm} denote the distributions of the pairs (ξ', η'), (ξ'', η''), $(\xi' + \xi'', \eta' + \eta'')$, respectively. The numbers p_{lm} are, in general, not completely defined by specifying the numbers p_{lm}' and p_{lm}''; for this it is also necessary to know the dependence between the pairs (ξ', η') and (ξ'', η''). In the most important case, when the latter pairs are mutually independent (i.e., when the values taken by the variables ξ', η' do not depend on the law p_{lm}'', and conversely) we easily obtain the rule of composition expressing the numbers p_{lm} in terms of the numbers p_{lm}' and p_{lm}''. Moreover, we can obtain the rule of composition for the addition of an arbitrary number of (mutually independent) pairs. These formulas (which we shall not introduce here) are completely analogous to formulas (1) and (2), which were established above for the one-dimensional case, but are, of course, substantially more complicated than (1) and (2).

§2. Limit theorems

In the theory of probability, as in every mathematical theory of a natural science, such as theoretical mechanics, thermodynamics and many others, one tries to establish conformance to the most general principles. These principles would relate not only to the particular processes taking place in nature and in human practice, but would include the widest possible class of phenomena. For instance, the fundamental theorems of mechanics — the theorem of kinetic energy, the (Keplerian) theorem of areas, etc. — are not related to any special form of mechanical motion, but to an extremely wide class of such motions. In the same way, the fundamental propositions of the theory of probability (such as the law of large numbers) not only include special forms of mass phenomena, but include extremely wide classes of them. It can be said that the essence of mass phenomena is revealed in regularities of this type, i.e., those properties of these phenomena are revealed which are due to their mass character, but which depend in only a relatively slight manner on the individual nature of the objects composing the masses. For example, sums of random variables, the individual terms of which may be distributed according to any of a wide range of laws, obey the law of large numbers; neither the applicability nor the content of the law of large numbers depends upon these individual distribution laws (which must satisfy only certain very general requirements).

One of the most important parts of the theory of probability — the theory of limit theorems — was developed because of this desire to establish general principles of a type which includes the widest possible class of real phenomena. In a very large number of cases — in particular, in the simplest problems which arose initially — the mass character of the phenomenon being studied was taken into account mathematically by investigating sums of a very large number of random variables (more or less equally significant and mutually dependent or independent). Thus, in the theory of measurement errors (one of the first applications of the theory of probability) we study the error actually incurred in performing a measurement; this error is usually the sum of a large number of individual errors caused by very different factors. The law of large numbers is concerned with just such sums of a large number of random variables. In the XVIIIth century, De Moivre and Laplace showed that in some of the simplest cases the sums of large numbers of mutually independent random variables, after proper normalization, were subject to distribution laws which approach the so-called "normal" law as the number of terms approaches infinity. The "density" of this law is given by the function

$$(2\pi)^{-\frac{1}{2}}e^{-\frac{1}{2}x^2}.$$

This was the first of the limit theorems of the theory of probability, the so-called theorem of De Moivre and Laplace. It is now studied in all courses on the theory of probability. This theorem includes only an extremely narrow class of cases, the so-called Bernoulli trials, where each term has as its possible values only the numbers 0 and 1, and the probabilities of these values are the same for all terms. However, as was stated by Laplace, the causes, due to which distribution laws of sums in the case of Bernoulli trials have a tendency to approach the normal law, have a character so general that there is every reason to suppose that the theorem of De Moivre and Laplace is merely a special case of some much more general principle. Laplace attempted to find the basis for this tendency to the normal law for a wider class of situations. However, neither he nor his contemporaries made significant progress in this direction, partly because the methods of mathematical analysis known at that time were inadequate for this purpose. The first method, by which it was possible to prove that the limit theorem is a general principle governing the behavior of sums of a large number of mutually independent random variables, was not formulated until the middle of the XIXth century by P. L. Chebyshev, the great Russian scholar. It is well-known that the first general conception of the law of large numbers is due to him. In general, the desire to establish principles of wide validity, which is common to every natural science and of which we spoke at the beginning of the present section, was noticeable in the theory of probability only after the investigations of Chebyshev.

Chebyshev tried to formulate a general limit theorem during almost all of his scientific life. He finally found a suitable formulation, but did not succeed in proving the theorem itself. The proof was completed shortly after Chebyshev's death by his student and successor A. A. Markov. However, several years before the work of Markov, A. M. Liapunov, who was also a student of Chebyshev, proved the limit theorem under extremely general conditions by a different method, which more closely resembles the contemporary proof.

The central limit theorem, first proved by Liapunov and later refined by other investigators, asserts that under certain general conditions the distribution law of a suitably normalized sum of a large number of mutually independent random variables should be close to the "normal" law introduced above. Here we are speaking of the so-called "integral" laws: If this normalized sum is denoted by S, then the probability of the inequality $S < x$ must be close to

$$(2\pi)^{-\frac{1}{2}} \int_{-\infty}^{x} e^{-\frac{1}{2}u^2} \, du.$$

It is immediately evident that in the general case, where the type of distribution laws of the summands themselves is unknown, any other formulation of the problem is impossible. Thus, for example, if the summands are integral-valued random variables, their sum may assume only integral values. To pose the question of the limiting behavior of the "density" of the distribution law of the sum would be meaningless in this case.

However, even though in the general case of arbitrarily distributed random variables only limit theorems of an "integral" type have meaning, it is still possible that if the terms of a given sum obey certain special types of laws, then "local" limit theorems may hold. These theorems can then give an approximate expression either for the probabilities of individually possible values of this sum or for the density of its distribution law, depending upon whether the summands obey discrete (in particular, integral-valued) or continuous distribution laws. Local limit theorems of both types turned out to be very important. Consequently, considerable attention has been given to their proof. For our purpose, the mathematical foundations of quantum statistics, only local limit theorems for the case of integral-valued random variables are necessary. Therefore, we consider only local limit theorems for integral-valued random variables and ignore the case of continuous distributions.

Suppose that we have an integral-valued random variable ξ subject to the distribution law p_k. The possible values of ξ are those integers k for which $p_k > 0$. The set of all pairwise differences of these possible values of the variable ξ has a greatest common divisor d. It is clear that if a_0 is one of the possible values of ξ, any arbitrary possible value of ξ may be represented in the form $a_0 + ld$, where l is an integer. The converse is in general not true: $a_0 + ld$ need not be a possible value of ξ for an arbitrary integer l. However, those numbers l for which $a_0 + ld$ is a possible value, evidently form a (finite or infinite) set of integers with the greatest common divisor 1.

Now let us assume that we have n mutually independent random variables $\xi_1, \xi_2, \cdots, \xi_n$, each subject to the distribution law described above. Then the sum s_n of these variables evidently assumes only values of the form $na_0 + ld$, where l is an integer. The problem of establishing a local limit theorem in this case consists in finding a suitable approximate expression for the probability

$$\mathbf{P}(s_n = na_0 + ld)$$

for different values of l.

In the modern theory of probability limit theorems (of both the integral and the local type) are extended to sums of random vectors. One tries to prove that under very general conditions the distribution laws of such sums, after a suitable normalization, approach the "normal" law in the proper number of dimensions when the number of summands approaches infinity. Assume, for example, that (ξ_i, η_i) $(i = 1, 2, \cdots, n)$ are mutually independent integral-valued random vectors obeying the same law

$$p_{ab} = \mathbf{P}(\xi_i = a, \eta_i = b).$$

Each pair of integers (a, b) represents a point in the plane having integral-valued Cartesian coordinates (a, b). Such points for which $p_{ab} > 0$ will be referred to as lattice points. These "possible" points of the vector (ξ_i, η_i) form a set of lattice points M in the plane. This set is a subset of the set of all lattice points of the plane. In general, it is possible (and in our applications it will often happen) that there exist in the plane other coarser parallelogram lattices of points covering the set M of possible points of our vector. The set of points of an arbitrary parallelogram lattice of the plane may be represented in the form

$$x = a_0 + k\alpha + l\beta \qquad (5)$$
$$y = b_0 + k\gamma + l\delta,$$

where $a_0, b_0, \alpha, \beta, \gamma, \delta$ are integral constants, $d = \alpha\delta - \beta\gamma \neq 0$, and k and l range over the complete set of integers. Every parallelogram whose vertices belong to this lattice and which contains no other points of the lattice in its interior or on its boundary is called a *fundamental parallelogram* of the lattice. The given lattice may have fundamental parallelograms of different form, but their area is always the same and is therefore a fundamental characteristic of the parallelogram lattice itself. This area, as may easily be calculated, is $|d| = |\alpha\delta - \beta\gamma|$. The parallelogram lattice covering the set M is called a *maximal* lattice if the area of its fundamental parallelogram has the largest possible value. This maximal area of a fundamental parallelogram plays the same role in the theory of two-dimensional random vectors as the number d, introduced above, plays in the one-dimensional case.

We shall call the integral-valued random vector (ξ, η) *degenerate* if all of its possible points lie on one straight line, i.e., if ξ and η are linearly dependent. We now prove that every non-degenerate vector has a maximal lattice. In fact, if the vector (ξ, η) is non-degenerate, then it has at least three possible points which do not lie on a straight line. We adjoin to them a fourth point so that a parallelogram P is obtained. Let the area of P be s. It is evident that P will be a parallelogram in any parallelogram lattice which covers the set M. Consequently, the fundamental parallelogram of such a lattice cannot have an area larger than s. It follows immediately that a maximal lattice can be found among such covering lattices [7].

If the distribution law of the random vector (ξ_i, η_i) $(i = 1, 2, \cdots, n)$ has the covering lattice (5), and if we assume that

$$\textstyle\sum_{i=1}^{n} \xi_i = S_n, \qquad \sum_{i=1}^{n} \eta_i = T_n,$$

then only points (x, y) of the form

$$x = na_0 + k\alpha + l\beta$$
$$y = nb_0 + k\gamma + l\delta,$$

where k and l are integers, can be possible points of the vector (S_n, T_n). Here the problem of establishing a local limit theorem consists in finding a convenient approximate expression for the probability

$$\mathbf{P}(S_n = na_0 + k\alpha + l\beta, T_n = nb_0 + k\gamma + l\delta)$$

for given integers k and l.

The remainder of this chapter will be devoted to establishing limit theorems of a similar type. However, in addition to finding the above-mentioned approximate expressions, we must also, for all cases of interest in subsequent applications, find an accurate estimate of the error caused by the replacement of the desired probability by an asymptotic expression.

In concluding the present section we make several historical remarks concerning the development of the study of limit theorems that followed the first discoveries of Liapunov and Markov. This development continued (and continues to the present time) in the following fundamental directions: 1) the extension of the limit theorems to multi-dimensional random variables (random vectors); 2) the proof of local limit theorems (of continuous and discrete types); 3) the extension of limit theorems to sums of mutually dependent random variables; 4) the search for the broadest possible (necessary and sufficient) conditions for the applicability of limit theorems; and 5) the refinement of the estimate of the remainder terms. In this exposition we limit ourselves to problems in which the limiting laws are normal. Of course, there is also the question of what other laws under what conditions can serve as limiting laws for sums of a large number of random variables. While this question has led to a particularly well-developed theory, we are not able to discuss it here [8].

§3. The method of characteristic functions

The most convenient method now known for proving the limit theorems of the theory of probability is the so-called method of characteristic functions. This is particularly so when the terms of the sum are mutually independent. We present the fundamentals of this method as applied to the case of interest to us, namely that of integral-valued random variables.

Let ξ be an integral-valued random variable obeying the law

$$p_n = \mathbf{P}(\xi = n) \qquad (-\infty < n < \infty).$$

If $f(\xi)$ is an arbitrary (real or complex) function of the variable ξ and if the series

$$\sum_{n=-\infty}^{\infty} f(n) p_n = \mathbf{E}f(\xi)$$

converges absolutely, then the sum of this series is what we agreed in §1 to call the mathematical expectation of the random variable $f(\xi)$. Let us assume, in particular, that

$$f(\xi) = e^{it\xi},$$

where $i = (-1)^{\frac{1}{2}}$ and t is an arbitrary real number. Since the series

$$\sum_{n=-\infty}^{\infty} e^{int} p_n$$

evidently converges absolutely for all real t, the mathematical expectation

$$\mathbf{E}e^{it\xi} = \sum_{n=-\infty}^{\infty} p_n e^{int}$$

of the complex random variable $e^{it\xi}$ exists for all real t and is a function of the real parameter t. We denote this function by $\varphi(t)$ and call it the *characteristic function* of the variable ξ or of the distribution law p_n. Thus,

$$\varphi(t) = \sum_{n=-\infty}^{\infty} p_n e^{int}. \tag{6}$$

We now mention some of the simplest general properties of characteristic functions.

1°. Evidently, we always have

$$\varphi(0) = \sum_{n=-\infty}^{\infty} p_n = 1,$$

and for real t

$$|\varphi(t)| \leq \sum_{n=-\infty}^{\infty} p_n = 1.$$

2°. If the variable ξ has the mathematical expectation $\mathbf{E}\xi$, then the latter can be represented by the series

$$\sum_{n=-\infty}^{\infty} np_n ,$$

which converges absolutely. In this case the series

$$\sum_{n=-\infty}^{\infty} inp_n e^{int},$$

which is obtained from the series (6) by means of termwise differentiation with respect to t, obviously converges absolutely and uniformly over the whole real line. Thus, the differentiability of the characteristic function $\varphi(t)$ for arbitrary t, and also the relation

$$\varphi'(0) = i\mathbf{E}\xi,$$

follow from the existence of the mathematical expectation $\mathbf{E}\xi$ of the variable ξ.

3°. In completely analogous fashion we easily convince ourselves that the existence of the kth moment of the variable ξ (where k is an arbitrary positive integer) implies the kth order differentiability of the function $\varphi(t)$, and also the relation

$$\varphi^{(k)}(0) = i^k\mathbf{E}\xi^k.$$

4°. From equation (6) it follows immediately that the characteristic function of an integral-valued random variable is always a periodic function, and that it always has a period (not necessarily the smallest) equal to 2π.

5°. Let ξ' and ξ'' be two mutually independent integral-valued random variables obeying, respectively, the distribution laws p_n' and p_n'', and having the characteristic functions $\varphi_1(t)$ and $\varphi_2(t)$, so that

$$\varphi_1(t) = \sum_{n=-\infty}^{\infty} p_n' e^{int}, \qquad \varphi_2(t) = \sum_{n=-\infty}^{\infty} p_n'' e^{int}.$$

Let p_n and $\varphi(t)$ denote, respectively, the distribution law and the characteristic function of the sum $\xi = \xi' + \xi''$. Then, as we have seen in §1,

$$p_n = \sum_{k=-\infty}^{\infty} p_k' p_{n-k}'',$$

and consequently

$$\begin{aligned}\varphi(t) &= \sum_{n=-\infty}^{\infty} p_n e^{int} = \sum_{n=-\infty}^{\infty} e^{int} \sum_{k=-\infty}^{\infty} p_k' p_{n-k}'' \\ &= \sum_{n=-\infty}^{\infty} \left(\sum_{k=-\infty}^{\infty} p_k' e^{ikt} p_{n-k}'' e^{i(n-k)t}\right) \\ &= \sum_{k=-\infty}^{\infty} p_k' e^{ikt} \sum_{n=-\infty}^{\infty} p_{n-k}'' e^{i(n-k)t} \\ &= \sum_{k=-\infty}^{\infty} p_k' e^{ikt} \sum_{l=-\infty}^{\infty} p_l'' e^{ilt} = \varphi_1(t)\varphi_2(t).\end{aligned}$$

If we considered as proved the theorem that the mathematical expectation of the product of mutually independent random variables is equal to the product of the individual expectations, then the result just obtained follows very simply from

$$\begin{aligned}\varphi(t) &= \mathbf{E}e^{it\xi} = \mathbf{E}e^{it(\xi'+\xi'')} \\ &= \mathbf{E}(e^{it\xi'}e^{it\xi''}) \\ &= \mathbf{E}e^{it\xi'}\mathbf{E}e^{it\xi''} = \varphi_1(t)\varphi_2(t).\end{aligned}$$

Thus, the characteristic function of the sum of mutually independent random variables is equal to the product of the individual characteristic functions. This rule of composition of characteristic functions was proved for the addition of only two random variables. Evidently, by using mathematical induction, the proof can be extended immediately to the case of the sum of an arbitrary number of (mutually independent) random variables.

This exceptional simplicity of the rule of composition of characteristic functions makes them an extremely convenient instrument for investigating sums of large numbers of mutually independent random variables, and, in particular, for proving limit theorems. On the other hand, the rule of composition of the distribution laws themselves, given by formulas (1) and (2) of §1, is quite complex, particularly for a large number of summands. For characteristic functions this rule, as we see, is distinguished by extraordinary simplicity, so that knowing the characteristic functions of the summands we can immediately and directly form the characteristic function of the sum.

However, for the characteristic functions to be a sufficiently powerful tool for the investigation of sums of a large number of random variables, one simple rule of composition is not enough. After finding the characteristic function of the sum, we should be able to establish, with its help, the distribution law of this sum. At present, we have related the distribution law to the corresponding characteristic function only through formula (6) which expresses $\varphi(t)$ in terms of p_n. We not only do not have the inverse relation which gives p_n in terms of $\varphi(t)$, but in essence we do not even know if the characteristic function $\varphi(t)$ defines the corresponding distribution p_n uniquely. All these questions, as we now show, are easily resolved by means of the classical formulas of Fourier.

6°. We multiply both sides of equation (6) by e^{-imt}, where m is an arbitrary integer, and integrate the expression obtained with respect to t from $-\pi$ to π. The series on the right side, being uniformly convergent in the range indicated, may be integrated termwise. We find

$$\int_{-\pi}^{\pi} e^{-imt}\varphi(t)\,dt = \sum_{n=-\infty}^{\infty} p_n \int_{-\pi}^{\pi} e^{i(n-m)t}\,dt,$$

or, since on the right side the integral is obviously equal to 2π for $n = m$ and to zero for $n \neq m$,

$$\int_{-\pi}^{\pi} e^{-imt}\varphi(t)\,dt = 2\pi p_m .$$

Hence,

$$p_m = (2\pi)^{-1} \int_{-\pi}^{\pi} e^{-imt}\varphi(t)\,dt. \tag{7}$$

This "inversion formula" shows that the distribution law of an integral-valued random variable is uniquely determined by its characteristic function $\varphi(t)$. At the same time formula (7) gives an extremely simple expression for this dependence.

7°. In §1, we wrote the whole set of possible values of the random variable ξ in the form $a_0 + ld$, where d is the greatest common divisor of the pairwise differences of these values, a_0 one of the possible values, and l an integer. In the sequel we agree to call the number d, which is uniquely determined by the distribution law of the variable ξ, the *increment* of this variable. If, as previously, we denote the distribution law of the variable ξ by p_n, then evidently $p_n > 0$ only if

$$n \equiv a_0 \pmod{d},$$

i.e., if the number n has the form $a_0 + ld$. Therefore, in contrast to what was done previously, it will be convenient to denote by p_l the probability that $\xi = a_0 + ld$. In this notation the characteristic function of the variable ξ is expressed by the formula

$$\varphi(t) = \sum_{l=-\infty}^{\infty} p_l e^{i(a_0+ld)t} = e^{ia_0t} \sum_{l=-\infty}^{\infty} p_l e^{ildt}.$$

We immediately see that the function

$$\varphi(t)e^{-ia_0t} = \sum_{l=-\infty}^{\infty} e^{ildt} p_l$$

has period $2\pi/d$. A calculation carried out in analogy with that of 6° easily shows that

$$p_l = (d/2\pi) \int_{-\pi/d}^{\pi/d} e^{-i(a_0+ld)t}\varphi(t)\,dt. \tag{8}$$

The great value of the concept of the increment d of an integral-valued random variable is explained by the following lemma:

LEMMA 1. *Let ξ be an integral-valued random variable having increment d and characteristic function $\varphi(t)$. Then, $|\varphi(t)| < 1$ for $0 < t \leq \pi/d$.* (The presence of the increment indicates that the random variable ξ is non-degenerate, i.e., that it has no less than two possible values.)

Indeed, if for an arbitrary value t we have

$$|\varphi(t)| = 1,$$

then

$$\left|\sum_{l=-\infty}^{\infty} p_l e^{ildt}\right| = |\varphi(t)e^{-ia_0t}| = 1.$$

Since $\sum_{l=-\infty}^{\infty} p_l = 1$, the above is possible only if all members of the sum for which $p_l > 0$ have the same argument θ, i.e., if $p_l > 0$ implies that

$$ldt = \theta + 2\pi u_l,$$

where u_l is an integer. Suppose $p_{l'} > 0$ and $p_{l''} > 0$. Then

$$(l'' - l')\,d = (u_{l''} - u_{l'})(2\pi/t).$$

But $(l'' - l')d = (a_0 + l''\,d) - (a_0 + l'\,d)$ is the difference between any two possible values of ξ. Thus, all such differences are integral multiples of $2\pi/t$. This means that the greatest common divisor of these differences, by definition equal to d, is an integral multiple of $2\pi/t$. Hence,

$$d = s(2\pi/t),$$

where s is an integer. Therefore,

$$t = 2\pi s/d.$$

Since the region $0 < |t| \leq \pi/d$ does not contain such values of t, it follows that $|\varphi(t)| < 1$ in this region, which was to be proved.

The method of characteristic functions can be generalized in a natural fashion to treat multi-dimensional random variables (random vectors). Let (ξ, η) be an integral-valued random vector subject to the distribution law

$$p_{ab} = \mathbf{P}(\xi = a, \eta = b).$$

The mathematical expectation of the variable $e^{it\xi+is\eta}$ is equal to

$$\mathbf{E}e^{i(t\xi+s\eta)} = \sum_{a,b=-\infty}^{\infty} e^{i(ta+sb)}p_{ab},$$

where t and s are real parameters. We will call this the characteristic function of the vector (ξ, η) or of the law p_{ab}, and designate it by $\varphi(t, s)$, so that

$$\varphi(t, s) = \sum_{a,b=-\infty}^{\infty} p_{ab}e^{i(ta+sb)}.$$

As in the one-dimensional case, the function $\varphi(t, s)$ is periodic with period 2π relative to each of the two parameters; $\varphi(0, 0) = 1$, and for any real t, s, we have $|\varphi(t, s)| \leq 1$.

Further, if the moment $\mathbf{E}(\xi^h\eta^k)$ exists, then the partial derivative

$$\partial^{h+k}\varphi/\partial t^h \partial s^k$$

also exists, and the value of this derivative for $t = s = 0$ is equal to

$$i^{h+k}\mathbf{E}(\xi^h\eta^k).$$

If (ξ_1, η_1) and (ξ_2, η_2) are two mutually independent integral-valued random vectors with the characteristic functions $\varphi_1(t, s)$ and $\varphi_2(t, s)$ and if $\varphi(t, s)$ is the characteristic function of the vector $(\xi_1 + \xi_2, \eta_1 + \eta_2)$, then

$$\varphi(t, s) = \varphi_1(t, s)\varphi_2(t, s).$$

This rule of composition remains valid for the addition of an arbitrary number of mutually independent random vectors.

Further, the "inversion formula"

$$p_{ab} = (1/4\pi^2) \int_{-\pi}^{\pi} \int_{-\pi}^{\pi} e^{-i(ta+sb)}\varphi(t, s)\, dt\, ds$$

holds.

All of these results are established in complete analogy with the one-dimensional case and with the same ease.

Now let the law p_{ab} have the maximal lattice

$$\begin{aligned} a &= a_0 + k\alpha + l\beta, \\ b &= b_0 + k\gamma + l\delta, \end{aligned} \tag{9}$$

where $d = \alpha\delta - \beta\gamma \neq 0$. Then p_{ab} can be different from zero only if a and b have the form (9) for some integers k and l. Therefore, as in the one-dimensional case, we change the notation somewhat and set

$$p_{kl} = \mathbf{P}(\xi = a_0 + k\alpha + l\beta,\ \eta = b_0 + k\gamma + l\delta).$$

For the characteristic function of the vector (ξ, η) we thus obtain the expression

$$\begin{aligned} \varphi(t, s) &= \textstyle\sum_{k,l=-\infty}^{\infty} p_{kl}e^{i[t(a_0+k\alpha+l\beta)+s(b_0+k\gamma+l\delta)]} \\ &= e^{i(ta_0+sb_0)} \textstyle\sum_{k,l=-\infty}^{\infty} p_{kl}e^{i[t(k\alpha+l\beta)+s(k\gamma+l\delta)]}. \end{aligned} \tag{10}$$

Our next problem is to find the "inversion formula" expressing the law p_{kl} in terms of the function $\varphi(t, s)$. This will be analogous to formula (8) of the one-dimensional case. For this purpose, we choose an arbitrary pair of integers (k', l') and multiply both sides of (10) by

$$e^{-i[t(a_0+k'\alpha+l'\beta)+s(b_0+k'\gamma+l'\delta)]}.$$

We then take the integrals of both sides, extending them over the region D of the plane (t, s), characterized by the inequalities

$$\begin{aligned} -\pi &\leq \alpha t + \gamma s \leq \pi, \\ -\pi &\leq \beta t + \delta s \leq \pi, \end{aligned} \tag{D}$$

which evidently represent a parallelogram of area $4\pi^2/|d|$ with its center at the origin. We obtain

$$\iint_D e^{-i[t(a_0+k'\alpha+l'\beta)+s(b_0+k'\gamma+l'\delta)]}\varphi(t, s)\,dt\,ds$$
$$= \sum_{k,l=-\infty}^{\infty} p_{kl} \iint_D e^{i\{[(k-k')\alpha+(l-l')\beta]t+[(k-k')\gamma+(l-l')\delta]s\}}\,dt\,ds.$$

On the right side of this equation the integral corresponding to the values $k = k'$, $l = l'$ is evidently equal to the area of the region D, i.e., $4\pi^2/|d|$. We now show that all the remaining integrals on the right side vanish. We choose any one of these integrals and transform the variables in it by setting

$$\alpha t + \gamma s = u,$$
$$\beta t + \delta s = v.$$

Then the integral assumes the form

$$\int_{-\pi}^{\pi}\int_{-\pi}^{\pi} e^{i[(k-k')u+(l-l')v]}(|d|)^{-1}\,du\,dv$$
$$= (|d|)^{-1}\int_{-\pi}^{\pi} e^{i(k-k')u}\,du \int_{-\pi}^{\pi} e^{i(l-l')v}\,dv,$$

and clearly vanishes if at least one of the two differences $k - k'$, $l - l'$ is different from zero. Thus we find

$$\iint_D e^{-i[t(a_0+k'\alpha+l'\beta)+s(b_0+k'\gamma+l'\delta)]}\varphi(t, s)\,dt\,ds = (4\pi^2/|d|)p_{k'l'}.$$

Hence, dropping the primes over the letters, we obtain for an arbitrary pair of integers k, l,

$$(11)\qquad p_{kl} = (|d|/4\pi^2)\iint_D e^{-i[t(a_0+k\alpha+l\beta)+s(b_0+k\gamma+l\delta)]}\varphi(t, s)\,dt\,ds.$$

This is the inversion formula we shall need.

Finally, to prove the two-dimensional limit theorem, it is very important to establish the propostion analogous to Lemma 1: If the lattice (9) is maximal, then $|\varphi(t, s)| < 1$ everywhere within and on the boundary of the region D, except for the point $t = s = 0$.

Indeed, let (t, s) be an arbitrary point of the plane for which

$$|\varphi(t, s)| = 1.$$

Then, by (10),

(12) $$\left| \sum_{k,l=-\infty}^{\infty} p_{kl} e^{i[t(a_0+k\alpha+l\beta)+s(b_0+k\gamma+l\delta)]} \right| = 1.$$

Since

$$\sum_{k,l=-\infty}^{\infty} p_{kl} = 1,$$

all terms in the sum (12) in which $p_{kl} > 0$ must have the same argument (to within multiples of 2π). Thus, if $p_{kl} > 0$, then

$$(a_0 + k\alpha + l\beta)t + (b_0 + k\gamma + l\delta)s = \theta + 2m\pi,$$

where θ is a constant and m is an integer (depending on k and l). Hence, an arbitrary possible point (a, b) of the vector (ξ, η) satisfies the relation

$$at + bs = \theta + 2m\pi,$$

with m an integer.

Let us consider the points

$$t_1 = (2\pi/|d|)\,\delta, \qquad s_1 = -(2\pi/|d|)\beta,$$

$$t_2 = -(2\pi/|d|)\gamma, \qquad s_2 = (2\pi/|d|)\alpha$$

of the plane (t, s). It is clear that these points cannot be connected to the origin by the same straight line. It is easily verified that

$$|\varphi(t_1, s_1)| = |\varphi(t_2, s_2)| = 1.$$

Let us assume now that $|\varphi(t, s)| = 1$ at some point (t, s) different from the origin. As we just proved, an arbitrary possible point of the vector (ξ, η) then lies on one of the system of parallel lines

(13) $$xt + ys = \theta + 2m\pi,$$

where m is an integer. Since (t_1, s_1) and (t_2, s_2) lie on a straight line which does not intersect the origin, the straight line passing through the origin and the point (t, s) cannot contain both points (t_1, s_1) and (t_2, s_2). For definiteness, let (t_1, s_1) lie outside this line; then $t_1 s - s_1 t \neq 0$. Since $|\varphi(t_1, s_1)| = 1$, all possible points of the vector (ξ, η) belong, in addition to the system (13), to the system of straight lines

(14) $$xt_1 + ys_1 = \theta_1 + 2\pi n,$$

with n an integer. The points of intersection of the systems of straight lines (13) and (14) form a lattice which, as we have shown, covers the set M. An easy calculation shows that the area of the fundamental parallelogram of this lattice is

(15) $$4\pi^2/|s_1 t - t_1 s|.$$

If the point (t, s) lies, as we now assume, within or on the boundary of the region D, then by the definition of this region, $|\beta t + \delta s| \leq \pi$. Hence,

$$|s_1 t - t_1 s| = (2\pi/|d|)|\beta t + \delta s| \leq 2\pi^2/|d|.$$

It follows that the area (15) of the fundamental parallelogram of the lattice just constructed must be $\geq 2|d|$. This is impossible, since this lattice covers the set M for which $|d|$ is the area of the fundamental parallelogram of the *maximal* lattice (9). Thus, we have proved the following proposition:

LEMMA 2. *Suppose that all possible points (a, b) of the random vector (ξ, η) are covered by the maximal lattice*

$$a = a_0 + k\alpha + l\beta,$$

$$b = b_0 + k\gamma + l\delta.$$

Then $|\varphi(t, s)| < 1$ everywhere within and on the boundary of the region D defined by the inequalities

$$|\alpha t + \gamma s| \leq \pi,$$

$$|\beta t + \delta s| \leq \pi,$$

except at the point $t = s = 0$.

§4. The one-dimensional limit theorem

Let $\xi_1, \xi_2, \cdots, \xi_n$ be mutually independent integral-valued random variables subject to the same distribution law

$$p_l = \mathbf{P}(\xi_i = a_0 + ld) \qquad (-\infty < l < \infty),$$

where a_0 is one of the possible values and d is the increment of the variable ξ_i. Let us set

$$\xi_1 + \xi_2 + \cdots + \xi_n = s_n.$$

Evidently, the random variable s_n can assume only values of the form $na_0 + ld$ (l an integer), and for brevity we write

$$\mathbf{P}(s_n = na_0 + ld) = P_n(l).$$

Let

$$\varphi(t) = \sum_{l=-\infty}^{\infty} p_l e^{it(a_0+ld)}$$

be the characteristic function of ξ_i. Then, according to the rule of composition, the characteristic function of s_n is $[\varphi(t)]^n$, which we write briefly as $\varphi^n(t)$. Thus,

$$\sum_{l=-\infty}^{\infty} P_n(l) e^{it(na_0+ld)} = \varphi^n(t).$$

The inversion formula (8), §3 thus gives

$$(16)\quad P_n(l) = (d/2\pi)\int_{-\pi/d}^{\pi/d} e^{-it(na_0+ld)}\varphi^n(t)\,dt \qquad (-\infty < l < \infty).$$

Our aim is to find convenient approximate expressions for the distribution law $P_n(l)$ and to estimate the associated error. Formula (16) will serve as a starting point.

We assume that the distribution law p_l of ξ_i has finite moments up to the fifth order inclusive, i.e., that the series

$$\sum_{l=-\infty}^{\infty} |\, l\,|^5 p_l$$

converges. As we saw in §3, this implies the existence of the derivatives of $\varphi(t)$ up to the fifth order inclusive. In particular, the mathematical expectation

$$\mathbf{E}\xi_i = \sum_{l=-\infty}^{\infty} (a_0 + ld)p_l = -i\varphi'(0) \equiv a$$

and the dispersion

$$\mathbf{D}\xi_i = \mathbf{E}\xi_i^2 - (\mathbf{E}\xi_i)^2 = \sum_{l=-\infty}^{\infty} (a_0 + ld - a)^2 p_l$$
$$= -\varphi''(0) + [\varphi'(0)]^2 \equiv b$$

of ξ_i exist.

Let us set $na_0 + ld - na = u$. (For given l the variable u represents the deviation of the value $na_0 + ld$ of the random variable s_n from its mathematical expectation, which is evidently equal to na.) In the present section we show that for $n \to \infty$,

$$(17)\quad P_n(l) = d(2\pi nb)^{-\frac{1}{2}}e^{-u^2/2nb} + n^{-\frac{3}{2}}(m_0 + m_1 u) + O[n^{-\frac{5}{2}}(n^{\frac{1}{2}} + |\, u\,|^3)],$$

where m_0 and m_1 are constants (independent of n and u). Formula (17) is the form of the one-dimensional local limit theorem which we need. In the case of deviations u which are not too large, this formula gives a very accurate and at the same time very simple approximate estimate of the probability $P_n(l)$.

To prove (17) we first establish an auxiliary proposition, a refinement of Lemma 1, §3.

LEMMA 3. *For* $-\pi/d \le t \le \pi/d$ *we have*

$$|\,\varphi(t)\,| \le e^{-ct^2},$$

where c is a positive constant.

Proof. In virtue of our assumptions concerning the existence of the moments of the variable ξ_i, the function $\varphi(t)$ in some neighborhood of zero may be represented in the form

$$\varphi(t) = 1 + iat - \tfrac{1}{2}qt^2 + O(\,|\,t\,|^3),$$

where $a = \mathbf{E}\xi$, $q = \mathbf{E}\xi^2$. (For brevity, we omit the subscript from ξ.) Hence

$$\begin{aligned} |\varphi(t)|^2 &= (1 - \tfrac{1}{2}qt^2)^2 + a^2t^2 + O(|t|^3) \\ &= 1 - t^2(q - a^2) + O(|t|^3) \\ &= 1 - bt^2 + O(|t|^3), \end{aligned}$$

where $b = q - a^2 > 0$ is the dispersion of ξ. Hence,

$$|\varphi(t)|^2 \leq 1 - bt^2 + c_1|t|^3,$$

with c_1 a positive constant. Let us set $b/2c_1 = \sigma$. For $|t| < \sigma$ we obtain

$$\begin{aligned} |\varphi(t)|^2 &\leq 1 - bt^2 + c_1\sigma t^2 = 1 - \tfrac{1}{2}bt^2 \leq e^{-\frac{1}{2}bt^2}, \\ |\varphi(t)| &\leq e^{-bt^2/4}. \end{aligned} \tag{18}$$

But, for $\sigma \leq |t| \leq \pi/d$ we have, by Lemma 1, $|\varphi(t)| < 1$. Since the function $|\varphi(t)|$ is continuous, a constant $\mu < 1$ can be found such that

$$|\varphi(t)| < \mu$$

for $\sigma \leq |t| \leq \pi/d$. Determining the positive number c_2 from the equality

$$\mu = e^{-c_2(\pi/d)^2},$$

we have for $\sigma \leq |t| \leq \pi/d$,

$$|\varphi(t)| < \mu = e^{-c_2(\pi/d)^2} < e^{-c_2t^2}. \tag{19}$$

Now if c denotes the smaller of the numbers $b/4$ and c_2, then, in view of (18) and (19), we have in the whole interval $(-\pi/d, +\pi/d)$,

$$|\varphi(t)| \leq e^{-ct^2}.$$

This proves Lemma 3.

We now turn to the proof of (17). We divide this proof into steps so that it can be more easily comprehended.

1. Let us choose an arbitrary number λ, $\frac{3}{7} < \lambda < \frac{1}{2}$, which we subsequently consider to be constant, and let us put $n^{-\lambda} = \sigma$, so that the quantity σ approaches zero as $n \to \infty$. For constant $r \geq 0$, $a > 0$ and for $n \to \infty$,

$$\begin{aligned} \int_\sigma^\infty t^r e^{-ant^2}\,dt &= (an)^{-\frac{1}{2}(r+1)} \int_{a^{\frac{1}{2}}n^{\frac{1}{2}-\lambda}}^\infty u^r e^{-u^2}\,du \\ &< (an)^{-\frac{1}{2}(r+1)} e^{-\frac{1}{2}an^{1-2\lambda}} \int_0^\infty u^r e^{-\frac{1}{2}u^2}\,du = o(n^{-\omega}), \end{aligned} \tag{20}$$

where ω is any positive constant. (The above follows since

$$e^{-u^2} = e^{-u^2/2}e^{-u^2/2} \leq e^{-\frac{1}{2}an^{1-2\lambda}}e^{-u^2/2}$$

in the interval of integration.) Hence, by Lemma 3,

$$\left|\int_{\sigma<|t|<\pi/d} e^{-it(na_0+ld)}\varphi^n(t)\,dt\right| \le \int_{\sigma<|t|<\pi/d} |\varphi(t)|^n\,dt$$

$$\le 2\int_\sigma^\infty e^{-nct^2}\,dt = o(n^{-\omega});$$

and consequently, by (16),

$$(21)\quad P_n(l) = (d/2\pi)\int_{-\sigma}^{\sigma} e^{-it(na_0+ld)}\varphi^n(t)\,dt + o(n^{-\omega}) \quad (-\infty < l < \infty).$$

2. Formula (21) allows us to limit ourselves to the analysis of values of t between $-\sigma$ and $+\sigma$. In particular, the product $n\,|\,t\,|^3 < n^{1-3\lambda}$ can always be considered arbitrarily small because $\lambda > \frac{1}{3}$.

Since, according to our assumption, $\varphi(t)$ has a derivative of the fifth order at $t = 0$,

$$\varphi(t) = 1 + it\mathbf{E}\xi - \tfrac{1}{2}t^2\mathbf{E}\xi^2 - \tfrac{1}{6}it^3\mathbf{E}\xi^3 + \tfrac{1}{24}t^4\mathbf{E}\xi^4 + O(\,|\,t\,|^5).$$

Therefore, because $\mathbf{E}\xi = a$, $\mathbf{E}\xi^2 = b + a^2$,

$$\ln \varphi(t) = ita - \tfrac{1}{2}bt^2 + ic_3t^3 + c_4t^4 + O(\,|\,t\,|^5),$$

where c_3 and c_4 are real constants. Hence,

$$(22)\quad \begin{aligned} &-it(na+u) + n\ln\varphi(t) \\ &\qquad = -itu - \tfrac{1}{2}nbt^2 + ic_3nt^3 + c_4nt^4 + O(n\,|\,t\,|^5), \end{aligned}$$

where all the terms on the right side, starting with the third, are arbitrarily small for $|\,t\,| < \sigma = n^{-\lambda}$.

Since $\lambda > \frac{3}{7}$, we have $n^4t^{12} = o(n\,|\,t\,|^5)$ and consequently

$$\cos(c_3nt^3) = 1 - \tfrac{1}{2}c_3^2n^2t^6 + O(n\,|\,t\,|^5).$$

Hence, in virtue of

$$e^{c_4nt^4} = 1 + c_4nt^4 + O(n^2t^8) = 1 + c_4nt^4 + O(n\,|\,t\,|^5),$$

we obtain from (22) the following relation:

$$\begin{aligned} &e^{-it(na+u)}\varphi^n(t) \\ &\quad = e^{-itu-\frac{1}{2}nbt^2}(\cos c_3nt^3 + i\sin c_3nt^3)\{1 + c_4nt^4 + O(n\,|\,t\,|^5)\} \\ &\quad = e^{-itu-\frac{1}{2}nbt^2}\{1 - \tfrac{1}{2}c_3^2n^2t^6 + O(n\,|\,t\,|^5) + i\sin c_3nt^3\} \\ &\qquad\qquad \cdot\{1 + c_4nt^4 + O(n\,|\,t\,|^5)\} \\ &\quad = e^{-itu-\frac{1}{2}nbt^2}\{1 + c_4nt^4 - \tfrac{1}{2}c_3^2n^2t^6 + i\sin c_3nt^3 - \tfrac{1}{2}c_3^2c_4n^3t^{10} \\ &\qquad\qquad + ic_4nt^4\sin c_3nt^3 + O(n\,|\,t\,|^5)\}. \end{aligned}$$

But, because $\lambda > \frac{3}{7} > \frac{1}{3}$,

$$\tfrac{1}{2}c_3^2 c_4 n^3 t^{10} = O(n^2 \mid t \mid^7),$$

and we find finally

$$e^{-it(na+u)}\varphi^n(t) = e^{-itu-\frac{1}{2}nbt^2}\{1 + c_4 nt^4 - \tfrac{1}{2}c_3^2 n^2 t^6 + i \sin c_3 nt^3 + O(n \mid t \mid^5) + O(n^2 \mid t \mid^7)\}. \tag{23}$$

Because $na + u = na_0 + ld$, the left side of the above equation is the integrand of the integral (21). Therefore, we have obtained the required asymptotic expression for the integrand.

3. Let us first estimate the integral

$$(d/2\pi)\int_{-\sigma}^{\sigma} e^{-itu-\frac{1}{2}nbt^2}\,dt$$

of the main term of this formula. Using the well-known formula of Laplace:

$$(1/2\pi)\int_{-\infty}^{\infty} e^{-itu-\frac{1}{2}nbt^2}\,dt = (2\pi nb)^{-\frac{1}{2}}e^{-u^2/2nb}, \tag{L}$$

we obtain

$$(d/2\pi)\int_{-\sigma}^{\sigma} e^{-itu-\frac{1}{2}nbt^2}\,dt = d(2\pi nb)^{-\frac{1}{2}}e^{-u^2/2nb} - (d/2\pi)\int_{|t|>\sigma} e^{-itu-\frac{1}{2}nbt^2}\,dt.$$

Hence, by (20),

$$(d/2\pi)\int_{-\sigma}^{\sigma} e^{-itu-\frac{1}{2}nbt^2}\,dt = d(2\pi nb)^{-\frac{1}{2}}e^{-u^2/2nb} + o(n^{-\omega}), \tag{24}$$

where ω is any real number.

4. Now let us turn to the estimate of the remainder terms. First, we replace e^{-itu} on the right side of (23) by $\cos ut - i \sin ut$ and perform a term-by-term multiplication, putting $\cos ut = 1 + O(u^2t^2)$. We retain only even terms in t in the product since the odd terms vanish after integration. This yields for the set of remainder terms on the right side of (23) the expression

$$\begin{aligned} &e^{-\frac{1}{2}nbt^2}\{c_4 nt^4 - \tfrac{1}{2}c_3^2 n^2 t^6 + \sin ut \sin c_3 nt^3 \\ &\qquad + O(nu^2t^6) + O(n^2u^2t^8) + O(n \mid t \mid^5) + O(n^2 \mid t \mid^7)\} \\ &= e^{-\frac{1}{2}nbt^2}\{c_4 nt^4 - \tfrac{1}{2}c_3^2 n^2 t^6 + uc_3 nt^4 + O(\mid u \mid n^3 t^{10}) \\ &\qquad + O(\mid u \mid^3 nt^6) + O(u^2 nt^6) + O(u^2 n^2 t^8) \\ &\qquad + O(n \mid t \mid^5) + O(n^2 \mid t \mid^7)\}. \end{aligned} \tag{25}$$

Now let us integrate. We integrate between the limits $(-\infty, \infty)$ in all cases. In view of (20), the extension of the region of integration introduces an error only of order $o(n^{-\omega})$ for the first three terms (where ω is any real number) and only strengthens the estimates for the remaining terms. Since for any $r \geq 0$,

$$\int_{-\infty}^{\infty} |t|^r e^{-\frac{1}{2}nbt^2}\, dt = n^{-\frac{1}{2}(r+1)}\beta_r, \tag{26}$$

where β_r is a positive constant, integration of the first three terms of the expansion (25) yields an expression of the form

$$n^{-\frac{3}{2}}(m_0 + m_1 u),$$

with m_0 and m_1 independent of n and u.

After integration the remaining six terms of (25) yield, respectively,

$$O(|u|n^{-\frac{3}{2}}), \qquad O(|u|^3 n^{-\frac{5}{2}}), \qquad O(u^2 n^{-\frac{3}{2}}),$$
$$O(u^2 n^{-\frac{5}{2}}), \qquad O(n^{-2}), \qquad O(n^{-2}).$$

Therefore, the sum of all the integrals obtained can be written as

$$n^{-\frac{3}{2}}(m_0 + m_1 u) + O[n^{-\frac{5}{2}}(n^{\frac{1}{2}} + |u|^3)],$$

with m_0 and m_1 constants independent of n and u. The integral of the set of remainder terms taken over the interval $(-\sigma, +\sigma)$ differs from this expression by a quantity of order $o(n^{-\omega})$. But since the integral of the principal term is given by (24), we find

$$(d/2\pi)\int_{-\sigma}^{\sigma} e^{-it(na_0+ld)}\varphi^n(t)\, dt$$
$$= d(2\pi nb)^{-\frac{1}{2}}e^{-u^2/2nb} + n^{-\frac{3}{2}}(m_0 + m_1 u) + O[n^{-\frac{5}{2}}(n^{\frac{1}{2}} + |u|^3)],$$

where $u = na_0 + ld - na$, and m_0, m_1 are independent of n and u. In virtue of (21) this yields

$$P_n(l) = d(2\pi nb)^{-\frac{1}{2}}e^{-u^2/2nb} + n^{-\frac{3}{2}}(m_0 + m_1 u) + O[n^{-\frac{5}{2}}(n^{\frac{1}{2}} + |u|^3)], \tag{27}$$

which we had intended to establish.

The one-dimensional limit theorem expressed by (27) yields an estimate for the probability $P_n(l)$ which is accurate enough for all the applications needed in quantum statistics. However, a cruder (but simpler) estimate, which we now establish, is sufficient in the majority of cases. Returning to (23), we again substitute $\cos tu - i \sin tu$ for e^{-itu}, but this time we do not put $\cos tu = 1 + O(t^2u^2)$. Instead, we simply note that $\cos tu$ is an even function of t whose absolute value does not exceed unity. As before, this yields the principal term

$$e^{-itu-\frac{1}{2}nbt^2}.$$

Writing only terms in even powers of t, we obtain for the set of remainder terms the expression

$$e^{-\frac{1}{2}nbt^2}\{c_4nt^4 \cos tu - \tfrac{1}{2}c_3^2n^2t^6 \cos tu + (\sin tu) \sin c_3nt^3 + O(n\,|\,t\,|^5) + O(n^2\,|\,t\,|^7)\}$$
$$= e^{-\frac{1}{2}nbt^2}\{O(nt^4) + O(n^2t^6) + O(\,|\,u\,|\,nt^4)\}.$$

After integration between the limits $(-\infty, \infty)$, the three terms obtained yield, in view of (26),

$$O(n^{-\frac{3}{2}}), \qquad O(n^{-\frac{3}{2}}), \qquad O(\,|\,u\,|\,n^{-\frac{3}{2}}).$$

Consequently, as before, we find

$$(28) \qquad P_n(l) = d(2\pi nb)^{-\frac{1}{2}}e^{-u^2/2nb} + O[n^{-\frac{3}{2}}(1 + |\,u\,|\,)].$$

This is the cruder estimate of the probability $P_n(l)$ which we shall subsequently need.

In conclusion, we give a complete formulation of the one-dimensional limit theorem just proved.

THEOREM 1. *Let*

$$p_l = \mathbf{P}(\xi = a_0 + ld)$$

be the distribution law of the integral-valued random variable ξ *with a possible value* a_0, *increment* d, *mathematical expectation* a, *and dispersion* b. *Let* s_n *be the sum of* n *mutually independent random variables each subject to the law* p_l *and let*

$$\mathbf{P}(s_n = na_0 + ld) = P_n(l).$$

If the law p_l *has finite moments up to the fifth order inclusive, then as* $n \to \infty$ *the following relations hold uniformly in* l:

$$(28) \qquad P_n(l) = d(2\pi nb)^{-\frac{1}{2}}e^{-u^2/2nb} + O[n^{-\frac{3}{2}}(1 + |\,u\,|\,)]$$

and

$$(27) \quad P_n(l) = d(2\pi nb)^{-\frac{1}{2}}e^{-u^2/2nb} + n^{-\frac{3}{2}}(m_0 + m_1u) + O[n^{-\frac{5}{2}}(n^{\frac{1}{2}} + |\,u\,|^3)],$$

where $u = na_0 + ld - na$, *and* m_0, m_1 *are constants.*

§5. The two-dimensional limit theorem

Let $(\xi_1, \eta_1), (\xi_2, \eta_2), \cdots, (\xi_n, \eta_n)$ be mutually independent integral-valued random vectors subject to the same distribution law p_{kl} with the maximal lattice

$$a_0 + k\alpha + l\beta, \qquad b_0 + k\gamma + l\delta, \qquad \alpha\delta - \beta\gamma = d \neq 0,$$

so that

$$p_{kl} = \mathbf{P}(\xi_i = a_0 + k\alpha + l\beta,\ \eta_i = b_0 + k\gamma + l\delta)$$
$$(1 \le i \le n;\ -\infty < k < \infty;\ -\infty < l < \infty).$$

Let us put

$$\textstyle\sum_{i=1}^n \xi_i = S_n, \qquad \sum_{i=1}^n \eta_i = T_n.$$

Obviously, the only possible points for the random vector (S_n, T_n) are pairs of numbers of the form

$$S_n = na_0 + k\alpha + l\beta, \qquad T_n = nb_0 + k\gamma + l\delta, \tag{29}$$

which comprise a lattice of the vector (S_n, T_n) if k and l range through all the integers. For brevity, let us put

$$\mathbf{P}(S_n = na_0 + k\alpha + l\beta, T_n = nb_0 + k\gamma + l\delta) = P_n(k, l).$$

The characteristic function of the vector (ξ_i, η_i) is

$$\varphi(t, s) = \textstyle\sum_{k,l=-\infty}^{\infty} p_{kl} e^{i[(a_0+k\alpha+l\beta)t+(b_0+k\gamma+l\delta)s]}.$$

According to the composition rule, the characteristic function of the vector (S_n, T_n) is $[\varphi(t, s)]^n$ which we write briefly as $\varphi^n(t, s)$. Hence,

$$\textstyle\sum_{k,l=-\infty}^{\infty} P_n(k, l) e^{i[(na_0+k\alpha+l\beta)t+(nb_0+k\gamma+l\delta)s]} = \varphi^n(t, s).$$

Therefore, the inversion formula (11) of §3 yields

$$P_n(k, l) = (|d|/4\pi^2) \iint_D e^{-i[(na_0+k\alpha+l\beta)t+(nb_0+k\gamma+l\delta)s]} \varphi^n(t, s)\, dt\, ds, \tag{30}$$

where the region D is defined by the inequalities

$$|\alpha t + \gamma s| \le \pi, \qquad |\beta t + \delta s| \le \pi. \tag{D}$$

[We did not investigate the question of the maximality of the lattice (29) which we introduced for the vector (S_n, T_n). However, this is not necessary since the inversion formula (11) holds for any lattice, whether maximal or not, provided that the lattice includes all possible points of the vector.]

We now wish to establish a convenient approximate expression for the probability $P_n(k, l)$ and to find an accurate estimate of the error incurred in the approximation.

We assume that the initial distribution p_{kl} is not degenerate; hence, according to §2 it follows that the distribution has a maximal lattice. (This, however, we had already assumed at the beginning of this section.) Furthermore, as in the one-dimensional case, we assume that ξ_i and η_i have finite moments up to the fifth order inclusive, which is equivalent to assuming the convergence of the double series

$$\sum_{k,l=-\infty}^{\infty} (|k|^5 + |l|^5) p_{kl}.$$

The last assumption guarantees the existence of all the partial derivatives up to the fifth order inclusive of the characteristic function $\varphi(t, s)$ at the point $t = s = 0$. In particular, this guarantees the existence of the mathematical expectations

$$a_1 = \mathbf{E}\xi = -i(\partial\varphi/\partial t)_{t=s=0}, \qquad a_2 = \mathbf{E}\eta = -i(\partial\varphi/\partial s)_{t=s=0},$$

of the dispersions

$$b_{11} = \mathbf{E}[(\xi - a_1)^2] = [(\partial\varphi/\partial t)^2 - \partial^2\varphi/\partial t^2]_{t=s=0},$$
$$b_{22} = \mathbf{E}[(\eta - a_2)^2] = [(\partial\varphi/\partial s)^2 - \partial^2\varphi/\partial s^2]_{t=s=0},$$

and of the "mixed" second order moment

$$b_{12} = (b_{11}b_{22})^{\frac{1}{2}}R_{12} = \mathbf{E}[(\xi - a_1)(\eta - a_2)]$$
$$= [(\partial\varphi/\partial t)(\partial\varphi/\partial s) - \partial^2\varphi/\partial t\,\partial s]_{t=s=0}.$$

Hence, because of the assumed non-degeneracy of the vector (ξ_i, η_i), it follows from Schwarz's inequality that the correlation coefficient R_{12} of the quantities (ξ_i, η_i) is different from ± 1, i.e.,

$$|b_{12}| < (b_{11}b_{22})^{\frac{1}{2}},$$

or

$$\Delta = b_{11}b_{22} - b_{12}^2 > 0.$$

The mathematical expectations of the quantities S_n and T_n are na_1 and na_2, respectively. The deviations of the values $na_0 + k\alpha + l\beta$, $nb_0 + k\gamma + l\delta$ of these quantities from their mathematical expectations are therefore

$$na_0 + k\alpha + l\beta - na_1 = u_1, \qquad nb_0 + k\gamma + l\delta - na_2 = u_2,$$

respectively.

The purpose of the present section is to establish the validity as $n \to \infty$ of the following asymptotic expansion:

$$P_n(k, l) = |d|\,(2\pi n\Delta^{\frac{1}{2}})^{-1} e^{-(1/2\pi\Delta)\{b_{22}u_1^2 - 2b_{12}u_1u_2 + b_{11}u_2^2\}} \tag{31}$$
$$+ n^{-2}(m_0 + m_1u_1 + m_2u_2) + O[n^{-3}(n^{\frac{1}{2}} + |u_1|^3 + |u_2|^3)],$$

where m_0, m_1, m_2 are constants (independent of n, u_1 and u_2). Formula (31) is the most accurate form of the two-dimensional limit theorem that we shall need.

First, we prove an auxiliary proposition which permits us to extend Lemma 3 of the preceding section to the two-dimensional case (and, simultaneously, to make Lemma 2 of §3 more precise).

LEMMA 4. *There exists a constant $c > 0$ such that for any point (t, s) of the region D,*

$$|\varphi(t, s)| \leq e^{-c(t^2+s^2)}.$$

Proof. Because of our hypotheses, the function $\varphi(t, s)$ can be represented in some neighborhood of the origin in the form

$$\varphi(t, s) = 1 + i(a_1 t + a_2 s) - \tfrac{1}{2}(a_{11}t^2 + 2a_{12}ts + a_{22}s^2) + O(|t|^3 + |s|^3),$$

where the numbers $a_1, a_2, a_{11}, a_{12}, a_{22}$ are first and second moments of the random vector (ξ_i, η_i). Whence,

$$\begin{aligned} |\varphi(t, s)|^2 &= [1 - \tfrac{1}{2}(a_{11}t^2 + 2a_{12}ts + a_{22}s^2)]^2 + (a_1 t + a_2 s)^2 \\ &\quad + O(|t|^3 + |s|^3) \\ &= 1 - (a_{11}t^2 + 2a_{12}ts + a_{22}s^2) + (a_1 t + a_2 s)^2 \\ &\quad + O(|t|^3 + |s|^3) \\ &= 1 - [(a_{11} - a_1^2)t^2 + 2(a_{12} - a_1 a_2)ts + (a_{22} - a_2^2)s^2] \\ &\quad + O(|t|^3 + |s|^3). \end{aligned}$$

Here $a_{11} - a_1^2 = b_{11}$, $a_{12} - a_1 a_2 = b_{12}$, $a_{22} - a_2^2 = b_{22}$ are the second order central moments of the vector (ξ_i, η_i) which we discussed above. Thus,

$$|\varphi(t, s)|^2 = 1 - (b_{11}t^2 + 2b_{12}ts + b_{22}s^2) + O(|t|^3 + |s|^3).$$

But, we saw above that $\Delta = b_{11}b_{22} - b_{12}^2 > 0$ [non-degeneracy of the vector (ξ_i, η_i)]. Therefore, the form $b_{11}t^2 + 2b_{12}ts + b_{22}s^2$ is positive definite (since $b_{11} > 0$, $b_{22} > 0$). This means that there exists a positive number c_1 such that

$$b_{11}t^2 + 2b_{12}ts + b_{22}s^2 \geq c_1(t^2 + s^2)$$

identically. Therefore, in some neighborhood of the origin,

$$|\varphi(t, s)|^2 \leq 1 - c_1(t^2 + s^2) + c_2(|t|^3 + |s|^3),$$

where c_2 is another positive constant. If $\sigma < c_1/2c_2$ and $|t| < \sigma$, $|s| < \sigma$, then

$$c_2(|t|^3 + |s|^3) \leq \tfrac{1}{2}c_1(t^2 + s^2).$$

Hence,

$$|\varphi(t, s)|^2 \leq 1 - \tfrac{1}{2}c_1(t^2 + s^2) \leq e^{-\frac{1}{2}c_1(t^2+s^2)},$$

or

(32) $$|\varphi(t, s)| \leq e^{-\frac{1}{4}c_1(t^2+s^2)} \qquad (|t| < \sigma, |s| < \sigma).$$

Let us denote by Q the square $|t| < \sigma$, $|s| < \sigma$ of the (t, s)-plane. In view of Lemma 2, we have $|\varphi(t, s)| < 1$ everywhere within the (closed) region $D - Q$. Since the function $|\varphi(t, s)|$ is continuous, a number $\mu < 1$ exists such that $|\varphi(t, s)| < \mu$ everywhere in the region $D - Q$. If we denote by Λ the largest value of the sum $t^2 + s^2$ in the region D and choose $c_3 > 0$ sufficiently small so that

$$\mu < e^{-c_3\Lambda},$$

then

$$|\varphi(t, s)| < \mu < e^{-c_3\Lambda} \leq e^{-c_3(t^2+s^2)} \tag{33}$$

everywhere in the region $D - Q$. Finally, if we denote by c the smaller of the numbers $\frac{1}{4}c_1$ and c_3, then in view of (32) and (33), we have

$$|\varphi(t, s)| \leq e^{-c(t^2+s^2)}$$

both in the region Q and in the region $D - Q$ and hence in the whole region D. This proves Lemma 4.

Turning now to the proof of the asymptotic formula (31), we again divide our discussion into steps corresponding to those of the preceding section in which we proved the one-dimensional limit theorem.

1. We again denote by λ an arbitrary constant lying between $\frac{3}{7}$ and $\frac{1}{2}$ and put $n^{-\lambda} = \sigma$.

The double integral in (30) extends over the region D defined by the inequalities (D). We divide this region into two parts, one being the square Q: $|t| \leq \sigma$, $|s| \leq \sigma$, and the other the complement $D - Q$ of the square Q in the original region D. In the region $D - Q$ at least one of the numbers $|t|$, $|s|$ exceeds σ and so by Lemma 4, $|\varphi(t, s)| \leq e^{-c\sigma^2}$. Hence we find (taking into account that the area of the region D is $4\pi^2/|d|$)

$$\left| \iint_{D-Q} e^{-i[(na_0+k\alpha+l\beta)t+(nb_0+k\gamma+l\delta)s]}\varphi^n(t, s)\, dt\, ds \right| \leq (4\pi^2/|d|)e^{-cn^{1-2\lambda}} = o(n^{-\omega}),$$

with ω any real number. In view of (30), this yields for $-\infty < k, l < \infty$,

$$P_n(k, l) = (|d|/4\pi^2) \iint_Q e^{-i[(na_0+k\alpha+l\beta)t+(nb_0+k\gamma+l\delta)s]}\varphi^n(t, s)\, dt\, ds + o(n^{-\omega}), \tag{34}$$

in complete analogy with (21) of the preceding section.

2. Since the function $\varphi(t, s)$ has partial derivatives up to the fifth order inclusive at $t = s = 0$, we can (by assuming that $|t| < \sigma$, $|s| < \sigma$) write the expansion

$$\ln \varphi(t, s) = i\bar{P}_1(t, s) - \tfrac{1}{2}\bar{P}_2(t, s) + i\bar{P}_3(t, s) + \bar{P}_4(t, s) + O(\tau^5),$$

where $\tau = |s| + |t|$ and $\bar{P}_r(t, s)$ denotes a homogeneous real polynomial of degree r in t and s. In particular,

$$\bar{P}_1(t, s) = a_1 t + a_2 s,$$
$$\bar{P}_2(t, s) = b_{11}t^2 + 2b_{12}ts + b_{22}s^2,$$

and for any r,

$$\bar{P}_r(t, s) = O(\tau^r).$$

Hence we obtain for the logarithm of the integrand in (34),

$$-i(na_1 + u_1)t - i(na_2 + u_2)s + n \ln \varphi(t, s) = -i(u_1 t + u_2 s)$$
$$- \tfrac{1}{2}n(b_{11}t^2 + 2b_{12}ts + b_{22}s^2) + in\bar{P}_3 + n\bar{P}_4 + O(n\tau^5).$$

Since $\lambda > \frac{3}{7}$, we have $n^4\tau^{12} = o(n\tau^5)$ and consequently

$$\cos n\bar{P}_3 = 1 - \tfrac{1}{2}n^2\bar{P}_3^2 + O(n^4\tau^{12}) = 1 - \tfrac{1}{2}n^2\bar{P}_3^2 + O(n\tau^5).$$

Since it also follows that

$$e^{n\bar{P}_4} = 1 + n\bar{P}_4 + O(n^2\tau^3) = 1 + n\bar{P}_4 + O(n\tau^5),$$

we find

$$e^{-i[(na_1+u_1)t+(na_2+u_2)s]}\varphi^n(t, s)$$
$$= e^{-i(u_1t+u_2s)-\frac{1}{2}n\bar{P}_2}[\cos n\bar{P}_3 + i \sin n\bar{P}_3][1 + n\bar{P}_4 + O(n\tau^5)]$$
$$= e^{-i(u_1t+u_2s)-\frac{1}{2}n\bar{P}_2}[1 - \tfrac{1}{2}n^2\bar{P}_3^2 + O(n\tau^5) + i \sin n\bar{P}_3][1 + n\bar{P}_4 + O(n\tau^5)$$
$$= e^{-i(u_1t+u_2s)-\frac{1}{2}n\bar{P}_2}[1 + n\bar{P}_4 - \tfrac{1}{2}n^2\bar{P}_3^2 + i \sin n\bar{P}_3 - \tfrac{1}{2}n^2\bar{P}_3^2 n\bar{P}_4$$
$$+ n\bar{P}^4 i \sin n\bar{P}_3 + O(n\tau^5)].$$

But here, since $\lambda > \frac{3}{7} > \frac{2}{5}$,

$$\tfrac{1}{2}n^2\bar{P}_3^2 n\bar{P}_4 = O(n^3\tau^{10}) = O(n\tau^5), \qquad n\bar{P}_4 \sin n\bar{P}_3 = O(n^2\tau^7),$$

and we find, finally,

$$e^{-i[(na_0+u_1)t+(na_2+u_2)s]}\varphi^n(t, s)$$
$$(35) \quad = e^{-i(u_1t+u_2s)-\frac{1}{2}n\bar{P}_2}[1 + n\bar{P}_4 - \tfrac{1}{2}n^2\bar{P}_3^2 - i \sin n\bar{P}_3$$
$$+ O(n\tau^5) + O(n^2\tau^7)].$$

This asymptotic expression for the integrand of (34) is completely analogous to (23) of the preceding section.

3. First let us estimate the integral

$$(|d|/4\pi^2) \iint_Q e^{-i(u_1t+u_2s)-\frac{1}{2}n\bar{P}_2}\, dt\, ds$$

of the main term of (35). Since

$$u_1t + u_2s = u_1[t + (b_{12}s/b_{11})] + s[u_2 - (b_{12}u_1/b_{11})]$$

and

$$\bar{P}_2(t, s) = b_{11}t^2 + 2b_{12}ts + b_{22}s^2 = b_{11}[t + (b_{12}s/b_{11})]^2 + \Delta s^2/b_{11}\,,$$

we have

$$(|\,d\,|/4\pi^2)\int_{-\infty}^{\infty}\int_{-\infty}^{\infty} e^{-i(u_1t+u_2s)-\frac{1}{2}n\bar{P}_2(t,s)}\,dt\,ds$$

$$= (|\,d\,|/4\pi^2)\int_{-\infty}^{\infty} e^{-is[u_2-(b_{12}u_1/b_{11})]-\frac{1}{2}n(\Delta s^2/b_{11})}\,ds$$

$$\cdot\int_{-\infty}^{\infty} e^{-iu_1[t+(b_{12}s/b_{11})]-\frac{1}{2}nb_{11}[t+(b_{12}s/b_{11})]^2}\,dt$$

$$= (|\,d\,|/2\pi)\int_{-\infty}^{\infty} e^{-is[u_2-(b_{12}u_1/b_{11})]-\frac{1}{2}n(\Delta s^2/b_{11})}\,ds$$

$$\cdot(1/2\pi)\int_{-\infty}^{\infty} e^{-iu_1v-\frac{1}{2}nb_{11}v^2}\,dv.$$

Now let us apply formula (L) of the preceding section to both of the integrals on the right side. This gives for the first integral

$$|\,d\,|\,b_{11}^{\frac{1}{2}}(2\pi n\Delta)^{-\frac{1}{2}}e^{-(1/2n\Delta)[u_2-(b_{12}u_1/b_{11})]^2b_{11}},$$

and for the second

$$(2\pi nb_{11})^{-\frac{1}{2}}e^{-u_1^2/2nb_{11}}.$$

Thus, we find

$$(|\,d\,|/4\pi^2)\int_{-\infty}^{\infty}\int_{-\infty}^{\infty} e^{-i(u_1t+u_2s)-\frac{1}{2}n\bar{P}_2(t,s)}\,dt\,ds$$

$$= (|\,d\,|/2\pi n\Delta^{\frac{1}{2}})\,e^{-(1/2n\Delta)(b_{11}u_2^2-2b_{12}u_1u_2+b_{22}u_1^2)}.$$

This is precisely the leading term of the right side of (31). However, we integrated over the whole plane ($-\infty < t, s < \infty$) instead of over the square Q and we must therefore estimate the error incurred. To do this we note, for example, that

$$\int_{\sigma}^{\infty} ds \int_{-\infty}^{\infty} e^{-\frac{1}{2}n\bar{P}_2}\,dt = \int_{\sigma}^{\infty} e^{-\frac{1}{2}n\Delta s^2/b_{11}}\,ds \int_{-\infty}^{\infty} e^{-\frac{1}{2}nb_{11}[t+(b_{12}s/b_{11})]^2}\,dt$$

$$= \int_{\sigma}^{\infty} e^{-n\Delta s^2/2b_{11}}\,ds \int_{-\infty}^{\infty} e^{-\frac{1}{2}nb_{11}u^2}\,du = o(n^{-\omega})$$

for all real ω since (20) of §4 gives this estimate for the first factor, and the second factor approaches zero as $n \to \infty$. Hence we can consider

$$(36)\quad (|d|/4\pi^2) \iint_D e^{-i(u_1t+u_2s)-\frac{1}{2}n\bar{P}_2(t,s)}\,dt\,ds$$
$$= (|d|/2\pi n\Delta^{\frac{1}{2}})\, e^{-(1/2n\Delta)(b_{11}u_2{}^2-2b_{12}u_1u_2+b_{22}u_1{}^2)} + o(n^{-\omega})$$

as established and can proceed to estimate the remaining terms.

4. For this purpose we replace $e^{-i(u_1t+u_2s)}$ in the right side of (35) by $\cos(u_1t + u_2s) - i\sin(u_1t + u_2s)$, in complete analogy with the computations of the previous section. Then we multiply term by term using the fact that $\cos(u_1t + u_2s) = 1 + O(u^2\tau^2)$, where $u = |u_1| + |u_2|$. As in the one-dimensional case, we retain only terms of even order in the variables t, s since the odd terms drop out after integration over the square Q. This gives for the set of remaining terms the expression

$$(37)\quad \begin{aligned} &e^{-\frac{1}{2}n\bar{P}_2}[n\bar{P}_4 - \tfrac{1}{2}n^2\bar{P}_3{}^2 - \sin n\bar{P}_3 \sin(u_1t + u_2s) \\ &+ O(u^2n^2\tau^8) + O(u^2n\tau^6) + O(n^2\tau^7) + O(n\tau^5)] \\ &\quad = e^{-\frac{1}{2}n\bar{P}_2}[n\bar{P}_4 - \tfrac{1}{2}n^2\bar{P}_3{}^2 - n\bar{P}_3(u_1t + u_2s) + O(un^3\tau^{10}) \\ &+ O(u^3n\tau^6) + O(u^2n^2\tau^8) + O(u^2n\tau^6) + O(n^2\tau^7) + O(n\tau^5)].\end{aligned}$$

Now we must integrate the expression (37) over the region Q. We can integrate the first three terms over the whole plane $(-\infty < t, s < \infty)$ instead of over the region Q and incur thereby an error only of order $n^{-\omega}$, where ω is any real number. In order to see this, it is evidently sufficient to show that

$$I(p, q) = \int_\sigma^\infty s^q\,ds \int_{-\infty}^\infty t^p e^{-\frac{1}{2}n\bar{P}_2}\,dt = o(n^{-\omega})$$

for all integers $p, q \geq 0$. But, putting $t + (b_{12}s/b_{11}) = z$, we evidently get

$$\bar{P}_2 = (\Delta s^2/b_{11}) + b_{11}z^2.$$

Hence,

$$I(p, q) = \int_\sigma^\infty s^q e^{-n\Delta s^2/2b_{11}}\,ds \int_{-\infty}^\infty [z - (b_{12}s/b_{11})]^p\, e^{-\frac{1}{2}nb_{11}z^2}\,dz.$$

Expanding $[z - (b_{12}s/b_{11})]^p$ by the binomial formula, we see (with no computation) that the integral $I(p, q)$ is represented as a sum of products of the form

$$\int_\sigma^\infty s^w e^{-n\Delta s^2/2b_{11}}\,ds \int_{-\infty}^\infty z^{w'} e^{-\frac{1}{2}nb_{11}z^2}\,dz,$$

where as n increases the first factor has the form $o(n^{-\omega})$ by (20) of the previous section, and the second factor in any case remains bounded. This shows that $I(u, v) = o(n^{-\omega})$.

In regard to the remaining terms, it is clear that extending the region of integration only strengthens the inequality obtained. Consequently, we can also integrate these terms over the whole (t, s)-plane instead of over the region Q.

For the homogeneous polynomial $\bar{P}_r(t, s)$ of degree r, the transformation $tn^{\frac{1}{2}} = t'$, $sn^{\frac{1}{2}} = s'$ yields

$$\int_{-\infty}^{\infty}\int_{-\infty}^{\infty} \bar{P}_r(t, s)e^{-\frac{1}{2}n\bar{P}_2(t,s)}\,dt\,ds = n^{-(\frac{1}{2}r+1)}c,$$

with c independent of n. Therefore, integrating the first three terms of the expansion (37) over the whole (t, s)-plane yields an expression of the form

$$n^{-2}(m_0 + m_1u_1 + m_2u_2),$$

where m_0, m_1, m_2 are independent of n, u_1 and u_2.

After integration we obtain the following estimates for the last six terms of (37):

$$O(un^{-3}),\ O(u^3n^{-3}),\ O(u^2n^{-3}),\ O(u^2n^{-3}),\ O(n^{-\frac{5}{2}}),\ O(n^{-\frac{5}{2}}).$$

Hence, after integration the whole set of remainder terms yields

$$n^{-2}(m_0 + m_1u_1 + m_2u_2) + O[n^{-3}(n^{\frac{1}{2}} + u^3)].$$

Combining this with the integral (36) of the main term, we find

$$\begin{aligned}(|\,d\,|/4\pi^2)\iint_Q e^{-i[(na_0+k\alpha+l\beta)t+(nb_0+k\gamma+l\delta)s]}\varphi^n(t, s)\,dt\,ds \\ = (\,|\,d\,|\,/2\pi n\Delta^{\frac{1}{2}})e^{-(1/2n\Delta)(b_{11}u_2^2-2b_{12}u_1u_2+b_{22}u_1^2)} \\ + n^{-2}(m_0 + m_1u_1 + m_2u_2) + O[n^{-3}(n^{\frac{1}{2}} + u^3)],\end{aligned}$$

where m_0, m_1, m_2 denote numbers independent of n, u_1 and u_2. In view of (34), we then have

$$\begin{aligned}P_n(k, l) = (\,|\,d\,|\,/2\pi n\Delta^{\frac{1}{2}})e^{-(1/2n\Delta)(b_{11}u_2^2-2b_{12}u_1u_2+b_{22}u_1^2)} \\ + n^{-2}(m_0 + m_1u_1 + m_2u_2) + O[n^{-3}(n^{\frac{1}{2}} + u^3)].\end{aligned} \tag{38}$$

This is the formula we wished to establish.

All that remains is for us to establish, in analogy with (28) of the preceding section, a cruder estimate for $P_n(k, l)$. This estimate will be completely adequate for many cases. In analogy with §4 we replace the factor $e^{-i(u_1t+u_2s)}$ in (35) by its trigonometric expression. We do not however replace cos $(u_1t + u_2s)$ by $1 + O(u^2\tau^2)$ as we did above. We simply use the fact that cos $(u_1t + u_2s)$ is an even function of t and s whose absolute value does not exceed unity. The remainder of the derivation is so closely analogous to the one-dimensional case that we need not repeat it here. We only cite the final result, which is

$$(39)\quad \begin{aligned} P_n(k, l) &= (\,|\,d\,|/2\pi n\Delta^{\frac{1}{2}})e^{-(1/2n\Delta)(b_{11}u_2^2 - 2b_{12}u_1u_2 + b_{22}u_1^2)} \\ &\quad + O[n^{-2}(1 + u)]. \end{aligned}$$

In conclusion, we give the complete formulation of the two-dimensional limit theorem just proved.

THEOREM 2. *Let*

$$p_{kl} = \mathbf{P}(\xi = a_0 + k\alpha + l\beta,\ \eta = b_0 + k\gamma + l\delta)$$

be the distribution of a non-degenerate integral-valued random vector (ξ, η) *with the maximal lattice* $a_0 + k\alpha + l\beta$, $b_0 + k\gamma + l\delta$, *the mathematical expectations* a_1, a_2, *the dispersions* b_{11}, b_{22}, *and the correlation coefficient* $b_{12}(b_{11}b_{22})^{-\frac{1}{2}}$. *Let* (S_n, T_n) *be the sum of* n *mutually independent random vectors each of which obeys the law* p_{kl} *and let*

$$P_n(k, l) = \mathbf{P}(S_n = na_0 + k\alpha + l\beta,\ T_n = nb_0 + k\gamma + l\delta).$$

If finite moments up to the fifth order inclusive exist for the law p_{kl}, *then as* $n \to \infty$ *the following relations hold uniformly in* k *and* l:

$$(39)\quad \begin{aligned} P_n(k, l) &= (\,|\,d\,|/2\pi n\Delta^{\frac{1}{2}})e^{-(1/2n\Delta)(b_{11}u_2^2 - 2b_{12}u_1u_2 + b_{22}u_1^2)} \\ &\quad + O[n^{-2}(1 + u)], \end{aligned}$$

and

$$(38)\quad \begin{aligned} P_n(k, l) &= (\,|\,d\,|/2\pi n\Delta^{\frac{1}{2}})e^{-(1/2n\Delta)(b_{11}u_2^2 - 2b_{12}u_1u_2 + b_{22}u_1^2)} \\ &\quad + n^{-2}(m_0 + m_1u_1 + m_2u_2) + O[n^{-3}(n^{\frac{1}{2}} + u^3)], \end{aligned}$$

where $d = \alpha\delta - \beta\gamma \neq 0$, $\Delta = b_{11}b_{22} - b_{12}^2 > 0$, $u_1 = na_0 + k\alpha + l\beta - na_1$, $u_2 = nb_0 + k\gamma + l\delta - na_2$, $u = |\,u_1\,| + |\,u_2\,|$, *and where* m_0, m_1, m_2 *denote constants (independent of* n, u_1 *and* u_2).

Chapter II

PRELIMINARY CONCEPTS OF QUANTUM MECHANICS

§1. Description of the state of a physical system in quantum mechanics

In classical mechanics the state of a system with s degrees of freedom is conveniently described by giving the values of all $2s$ "Hamiltonian variables": the "generalized coordinates" $q_1, q_2, \cdots, q_s$ and the "generalized momenta" $p_1, p_2, \cdots, p_s$. Thus, in classical mechanics the state of a system can be represented by a point in its "phase space", which is a Euclidean space of $2s$ dimensions with rectangular coordinates $q_1, q_2, \cdots, q_s, p_1, p_2, \cdots, p_s$. It is well-known that such a representation of the state of a system is particularly convenient for the purposes of statistical mechanics. Each physical quantity related to the system is a certain function $F(q_1, \cdots, q_s, p_1, \cdots, p_s)$ of its Hamiltonian variables with a definite value for each definite state of the system. In other words, such a quantity is a single-valued function defined on the whole phase space.

In quantum mechanics the problem of characterizing the state of a physical system is solved in an entirely different way. The state of the system is described by giving a certain (in general, complex) function

$$U = U(q_1, \cdots, q_s)$$

of the generalized coordinates of the system alone. (It is common practice to call such a function the "wave function" of the system.) A knowledge of the function U describing the state of the system does not in general make it possible to determine uniquely any of the Hamiltonian variables $q_1, \cdots, q_s, p_1, \cdots, p_s$. In the most favorable case, knowing the function U, we can determine the values of *some* of these variables, but *all* $2s$ variables can never be uniquely determined. The reason is that in quantum mechanics the quantities q_i and p_i (for the same value of i) cannot both have definite values in any single state. Of course it follows that any physical quantity $F(q_1, \cdots, q_s, p_1, \cdots, p_s)$ does not in general have a single definite value in the state described by the function U (or, as we say briefly, in the state U).

In quantum mechanics, prescribing the function U determines the *distribution laws* of physical quantities, but not their values. Thus, knowing the function U, we can calculate the probability that any physical quantity $\mathfrak{A}$ have a value included in a given interval (a, b), i.e., the probability of the inequality $a < \mathfrak{A} < b$. Only in the exceptional case, when the distribution law found for the quantity $\mathfrak{A}$ is such that $\mathfrak{A}$ has only one possible value, can it be said that $\mathfrak{A}$ has a definite value in the state U.

In particular, if the system is in a state U, the probability that the values of $q_1, \cdots, q_s$ belong to a given region V of the configuration space (i.e., the Euclidean space of s dimensions with rectangular coordinates $q_1, \cdots, q_s$) is given by

$$\int_V |U|^2 \, dq_1 \cdots dq_s \Big/ \int |U|^2 \, dq_1 \cdots dq_s ,$$

where the integral in the denominator is taken over the whole configuration space. In the even more special case of a system consisting of an elementary particle with the simplest Hamiltonian variables x, y, z, p_x, p_y, p_z, the probability that this particle belong to the region V of the ordinary three-dimensional space is expressed by

$$\int_V |U(x, y, z)|^2 \, dx \, dy \, dz \Big/ \int |U(x, y, z)|^2 \, dx \, dy \, dz ,$$

where the integral in the denominator is taken over the whole three-dimensional space.

Since the function U determines the distribution law of a physical quantity $\mathfrak{A}$, it is clear that U also determines uniquely the mathematical expectation $\mathsf{E}\mathfrak{A}$ of this quantity. In particular, if

$$\mathfrak{A} = F(q_1, \cdots, q_s)$$

is any function of the generalized coordinates $q_1, \cdots, q_s$, then

$$(1) \quad \mathsf{E}\mathfrak{A} = \int F(q_1, \cdots, q_s) |U|^2 \, dq_1 \cdots dq_s \Big/ \int |U|^2 \, dq_1 \cdots dq_s ,$$

where both integrals are taken over the whole configuration space.

All the formulas we have presented show that replacing the function U by a function λU, where $\lambda \neq 0$ is an arbitrary complex constant, does not produce any changes in the statistics of those physical quantities which depend only on the generalized coordinates $q_1, \cdots, q_s$ (i.e., that do not depend on the generalized momenta $p_1, \cdots, p_s$). We shall see later that this same rule also remains valid for physical quantities of any sort. Thus, we must regard the functions U and λU as describing the same state of the system, whatever may be the value of the complex constant $\lambda \neq 0$. Making use of this freedom in the choice of λ, we can obviously describe any state of the system by a wave function U for which

$$\int |U|^2 \, dq_1 \cdots dq_s = 1.$$

We then refer to the function U as *normalized*. For normalized functions U all the formulas we have written are simplified, since their denominators become unity. If the function U describing a state of the system is a normalized function and $\lambda = e^{i\varphi}$ is an arbitrary complex constant with absolute value 1, then λU is also normalized and describes the same state of the system.

It is obvious that the use of the function U to establish the statistics of physical quantities, in the way we have been describing, requires the convergence of all the integrals used in this process. In particular, $|U|^2$ must be integrable over the whole configuration space (that is, U must be an element of a complex Hilbert space). If this requirement is satisfied by U_1 and U_2, then the integral

$$\int U_1 U_2^* \, dq_1 \cdots dq_s$$

(where the star indicates complex conjugation) taken over the whole configuration space is absolutely convergent. We call this integral the *scalar* (*inner*) *product* of U_1 and U_2, and write it briefly as (U_1, U_2). In particular, it follows that

$$(U, U) = \int |U|^2 \, dq_1 \cdots dq_s .$$

Hence, if U is normalized,

$$(U, U) = 1.$$

Scalar products obviously have the following elementary properties:

1. $(U_1 + U_2, U_3) = (U_1, U_3) + (U_2, U_3),$

   $(U_1, U_2 + U_3) = (U_1, U_2) + (U_1, U_3).$

2. If λ is an arbitrary complex number, then

   $(\lambda U_1, U_2) = \lambda(U_1, U_2),$

   $(U_1, \lambda U_2) = \lambda^*(U_1, U_2).$

3. $(U_2, U_1) = (U_1, U_2)^*.$

One of the most important principles of quantum mechanics is the so-called *principle of superposition*: If the functions $U_1(q_1, \cdots, q_s)$ and $U_2(q_1, \cdots, q_s)$ describe possible states of the system, then the function $U = \lambda_1 U_1 + \lambda_2 U_2$, where λ_1, λ_2 are arbitrary complex constants and $|\lambda_1| + |\lambda_2| > 0$, describes a possible state of the system. This principle is automatically extended to the case of any number of components. It can also be extended to infinite series: If the series

$$\textstyle\sum_{k=1}^{\infty} \lambda_k U_k$$

converges in the mean to the function U [i.e., if $(U - S_n, U - S_n) \to 0$

as $n \to \infty$, where $S_n = \sum_{k=1}^{n} \lambda_k U_k]$, and if each of the functions U_k describes a possible state of the system, then so does U. In all cases the only exception is that of a sum identically equal to zero: The function $U \equiv 0$ does not describe any state of the system; it cannot be normalized and all the formulas we have given above for probabilities and mathematical expectations become meaningless for $U \equiv 0$.

A number of cases can be found in the history of statistical physics where an imprecisely defined concept of the probability of an event has led to confusion and to misunderstanding, and has thus hindered the development of science. In physics, as in all applied sciences, the probability of an event always means the relative frequency of its occurrence. The same event, however, can have very different probabilities under different conditions. Therefore, any assertion regarding probabilities has a precise meaning only when the conditions under which it applies are stated precisely.

In quantum mechanics, as we have already seen in part and shall see in considerably more detail below, the most important assertions are of the form: "If the system is in a state U and $\mathfrak{A}$ is a quantity related to this system, then the probability that $a < \mathfrak{A} < b$ is equal to the number p." To avoid confusion and ambiguity it is therefore necessary to first explain the meaning of such an assertion with complete precision. Suppose we have a large number n of systems of the type considered. Let all these systems exist in the state U and suppose that on each of these systems we carry out a measurement of the quantity $\mathfrak{A}$. We assume that when this is done the value of $\mathfrak{A}$ lies between the numbers a and b for m systems, and that for the other $n - m$ systems the value of $\mathfrak{A}$ lies outside the interval (a, b). The meaning of our probabilistic assertion is that *if n is a large number, the ratio m/n is close to p.*

This statistical meaning holds in general for all the probabilistic assertions of quantum mechanics. A very important feature of such a statistical prediction is that its only underlying assumption is the requirement that each of the n systems should be in the state U at the moment of measurement. All other conditions (which in various cases can be of very different sorts) have no influence at all on the probability of the event $a < \mathfrak{A} < b$. It is thus completely immaterial whether or not all the n measurements are carried out at the same place and at the same time or whether they are carried out at places extremely remote from each other both spatially and temporally. It is also completely immaterial when and in what manner each of the given systems has been brought into the state U. In particular, we can carry out all n measurements on a single system. All that is necessary is that after one measurement has been carried out the system must again be brought into the state U before the following measurement is made.

The fact that the statistical predictions of quantum mechanics are independent of all supplementary conditions is one of the most important general principles of the theory and gives the predictions a real statistical meaning.

§2. Physical quantities and self-adjoint linear operators

In §1 we saw the manner in which quantum mechanics allows us to define, for each state U, the statistics of a quantity $\mathfrak{A}$ which depends only on the generalized coordinates $q_1, \cdots, q_s$ of the system. But, in general, a physical quantity $\mathfrak{A}$ depends on the whole set of Hamiltonian variables $q_1, \cdots, q_s$, $p_1, \cdots, p_s$, and we now have to formulate the rules which will permit us to define the statistics of such a quantity for any state U of the system. For this purpose we use the concept of a linear self-adjoint operator.

An operator $\mathfrak{a}$ is a rule which assigns to each element U of the complex Hilbert space a definite element $\mathfrak{a}U$ of the same space. (Sometimes $\mathfrak{a}$ is not defined for all U but only for a subset of the set of all states.) An operator $\mathfrak{a}$ is said to be *linear* if

$$\mathfrak{a}(\lambda_1 U_1 + \lambda_2 U_2) = \lambda_1 \mathfrak{a} U_1 + \lambda_2 \mathfrak{a} U_2$$

for arbitrary U_1, U_2 and all complex constants λ_1, λ_2. To each operator $\mathfrak{a}$ there corresponds a definite complex adjoint operator $\mathfrak{a}^*$ defined by the condition

$$(\mathfrak{a}U_1, U_2) = (U_1, \mathfrak{a}^* U_2).$$

An operator $\mathfrak{a}$ is said to be *self-adjoint* (or *Hermitean*) if

$$(\mathfrak{a}U_1, U_2) = (U_1, \mathfrak{a}U_2) \tag{2}$$

for all U_1, U_2.

In quantum mechanics a linear self-adjoint operator $\mathfrak{a}$ is assigned to each physical quantity $\mathfrak{A}$. The statistics of the quantity are established for each state U of the system by means of this operator. The manner in which this is done will be stated below. At present, we note to begin with that if the quantity $\mathfrak{A}$ is a function of just the generalized coordinates $q_1, \cdots, q_s$, then the operator $\mathfrak{a}$ assigned to it is defined simply as multiplication by this function: If

$$\mathfrak{A} = F(q_1, \cdots, q_s),$$

then

$$\mathfrak{a}U(q_1, \cdots, q_s) = F(q_1, \cdots, q_s)U(q_1, \cdots, q_s).$$

The linearity of this operator is easily verified. Moreover, if the function F is real, then

$$(\mathcal{A}U_1, U_2) = \int (FU_1)U_2^* \, dq_1 \cdots dq_s$$

$$= \int U_1(FU_2)^* \, dq_1 \cdots dq_s = (U_1, \mathcal{A}U_2).$$

This shows that a real physical quantity of the form $\mathfrak{A} = F(q_1, \cdots, q_s)$ always has a definite self-adjoint operator corresponding to it.

We now consider the general case of quantities of the form

$$\mathfrak{A} = F(q_1, \cdots, q_s, p_1, \cdots, p_s).$$

The way in which the self-adjoint operator corresponding to such a quantity is defined will not interest us in this book. Therefore, we shall give only a brief description of it. First, we consider the simplest case $\mathfrak{A} = p_k$ $(1 \leq k \leq s)$. We choose the operator

$$\mathcal{P}_k = -i\hbar \, \partial/\partial q_k$$

to correspond to the generalized momentum p_k, i.e., we set

$$\mathcal{P}_k U = -i\hbar \, \partial U/\partial q_k,$$

where $\hbar$ is Planck's constant divided by 2π. This operator is clearly linear. To prove that it is self-adjoint we note that

$$(\mathcal{P}_k U_1, U_2) = (-i\hbar \, \partial U_1/\partial q_k, U_2)$$

$$= -i\hbar \int (\partial U_1/\partial q_k) U_2^* \, dq_1 \cdots dq_s.$$

But for fixed $q_1, \cdots, q_{k-1}, q_{k+1}, \cdots, q_s$,

$$\int_{-\infty}^{\infty} (\partial U_1/\partial q_k) U_2^* \, dq_k = U_1 U_2^* \Big|_{-\infty}^{\infty} - \int_{-\infty}^{\infty} U_1(\partial U_2^*/\partial q_k) \, dq_k$$

$$= -\int_{-\infty}^{\infty} U_1(\partial U_2^*/\partial q_k) \, dq_k$$

if, as we assume, the functions U_1, U_2 vanish for $q_k = -\infty$ and $q_k = \infty$. Therefore,

$$(\mathcal{P}_k U_1, U_2) = i\hbar \int U_1(\partial U_2^*/\partial q_k) \, dq_1 \cdots dq_s$$

$$= \int U_1(-i\hbar \, \partial U_2/\partial q_k)^* \, dq_1 \cdots dq_s$$

$$= (U_1, \mathcal{P}_k U_2),$$

as was to be proved. It is worthwhile pointing out that the self-adjoint property of the operator $\mathcal{P}_k$ depends in an essential way on the presence of the imaginary unit as a factor in the definition of this operator. The real operator $\hbar\, \partial/\partial q_k$ does not have this property of being self-adjoint, as can readily be verified.

If the physical quantity $\mathfrak{A}$ is an arbitrary function

$$\mathfrak{A} = F(q_1, \cdots, q_s, p_1, \cdots, p_s)$$

of the Hamiltonian variables of the system, the idea naturally arises of defining the corresponding operator $\mathcal{A}$ by means of the formula

$$\mathcal{A} = F(\mathcal{Q}_1, \cdots, \mathcal{Q}_s, \mathcal{P}_1, \cdots, \mathcal{P}_s),$$

where $\mathcal{Q}_k$, $\mathcal{P}_k$ denote, respectively, the operators assigned to the quantities q_k, p_k $(1 \leq k \leq s)$. This is indeed the path usually followed in quantum mechanics. This idea, which on the whole has been extremely fruitful, encounters a practical difficulty in regard to interpreting the operator

$$F(\mathcal{Q}_1, \cdots, \mathcal{Q}_s, \mathcal{P}_1, \cdots, \mathcal{P}_s). \tag{3}$$

The solution to this problem is by no means obvious even in the very simple case of F a polynomial with constant coefficients. To form the polynomial (3) it is necessary to define the procedures for adding and multiplying operators, i.e., it is necessary to formulate an algebra of operators. This algebra is constructed as follows: If $\mathcal{A}$ and $\mathcal{B}$ are operators, and α, β are real numbers, then the operator

$$\mathcal{C} = \alpha\mathcal{A} + \beta\mathcal{B}$$

is the operator defined by

$$\mathcal{C}U = \alpha\mathcal{A}U + \beta\mathcal{B}U.$$

If $\mathcal{A}$ and $\mathcal{B}$ are linear self-adjoint operators, $\mathcal{C}$ will be an operator of the same kind. We call the operator $\mathcal{C}$ defined by

$$\mathcal{C}U = \mathcal{A}(\mathcal{B}U)$$

the product $\mathcal{A}\mathcal{B}$ of the operators $\mathcal{A}$, $\mathcal{B}$. Hence the result of applying $\mathcal{A}\mathcal{B}$ is the same as the result of successive application to the same function of first $\mathcal{B}$ and then $\mathcal{A}$. Multiplication of operators is in general not commutative: $\mathcal{A}\mathcal{B}U$ is not in general the same as $\mathcal{B}\mathcal{A}U$. For example,

$$\mathcal{P}_k\mathcal{Q}_kU = -i\hbar\, \partial(Uq_k)/\partial q_k = -i\hbar[U + (q_k\, \partial U/\partial q_k)],$$

while

$$\mathcal{Q}_k\mathcal{P}_kU = q_k(-i\hbar\, \partial U/\partial q_k) = -i\hbar q_k\, \partial U/\partial q_k\, .$$

Due to the non-commutativity of multiplication for operators, the meaning of (3) is ambiguous, even in the case of a polynomial. This raises a difficulty in the choice of the operator corresponding to a given physical quantity. The product of two self-adjoint operators will be a self-adjoint operator only if the two factors commute with each other. We note also that since each operator obviously commutes with itself, a positive integral power of a self-adjoint operator is always a self-adjoint operator.

Thus we see that the problem of assigning an operator to a given physical quantity $\mathfrak{A}$ cannot always be solved by simple general methods. In many cases a special physical consideration is required. But, for the purposes of this book such difficulties are of no importance. Essentially, the only important fact for us is the possibility of assigning to each physical quantity a self-adjoint operator which will permit us to determine the statistics of the quantity in any state of the system in a unique and in principle very simple manner. We shall now consider the question of how these statistics are established once the operator for the given quantity is known.

When the physical quantity

$$\mathfrak{A} = F(q_1, \cdots, q_s)$$

depends only on the generalized coordinates $q_1, \cdots, q_s$, its mathematical expectation $\mathbf{E}\mathfrak{A}$ in the state U is given, as we know [§1, (1)], by

$$\mathbf{E}_U\mathfrak{A} = \int F(q_1, \cdots, q_s) \mid U(q_1, \cdots, q_s) \mid^2 dq_1 \cdots dq_s.$$

For simplicity, we have taken the function U to be normalized. Since the operator $\mathfrak{a}$ defined by

$$\mathfrak{a}U = F(q_1, \cdots, q_s)U$$

corresponds in this case to the quantity $\mathfrak{A}$, we may write

$$\mathbf{E}_U\mathfrak{A} = \int FUU^* \, dq_1 \cdots dq_s$$

$$= \int (\mathfrak{a}U)U^* \, dq_1 \cdots dq_s = (\mathfrak{a}U, U).$$

Extending this result, which we have established for quantities $\mathfrak{A}$ of a very special type, to the general case yields one of the basic principles of quantum mechanics: *If the linear self-adjoint operator $\mathfrak{a}$ corresponds to the physical quantity $\mathfrak{A}$, then the mathematical expectation of $\mathfrak{A}$ in the state U is equal to*

$$\mathbf{E}_U\mathfrak{A} = (\mathfrak{a}U, U). \tag{4}$$

[The function U is supposed normalized. In the general case obviously $\mathbf{E}_U\mathfrak{A} = (\mathfrak{a}U, U)/(U, U)$.]

This simple general rule contains the true meaning of the "assignment" to each physical quantity of a definite linear self-adjoint operator. We see that when we know the operator corresponding to a given quantity, we can actually determine by a single simple principle the mathematical expectation of the quantity in any state U.

It might be thought that the result we have found determines only the mathematical expectation of the quantity $\mathfrak{A}$, but does not determine its distribution law. This is actually true if we know only the operator corresponding to $\mathfrak{A}$. But in principle the method partly described above also makes it possible to determine the complete statistics of physical quantities related to the given system. Indeed, if we wish, for example, to find the probability of inequalities such as

$$a < \mathfrak{A} < b,$$

where a, b are real numbers; and if

$$\mathfrak{A} = F(q_1, \cdots, q_s, p_1, \cdots, p_s),$$

then it is sufficient to consider another physical quantity $\mathfrak{B}$ defined by

$$\mathfrak{B} = \begin{cases} 1 & \text{if } a < F < b, \\ 0 & \text{in all other cases.} \end{cases}$$

This quantity is obviously also a function of the variables $q_1, \cdots, q_s$, $p_1, \cdots, p_s$ and therefore can be regarded as a physical quantity related to the given system. On the other hand, it is obvious that the probability of the inequality $a < \mathfrak{A} < b$, when the system is in the state U, is equal to

$$\mathsf{P}_U(a < \mathfrak{A} < b) = \mathsf{E}_U\mathfrak{B}.$$

If the linear self-adjoint operator $\mathcal{B}$ corresponds to the quantity $\mathfrak{B}$, then we have by (4)

$$\mathsf{P}_U(a < \mathfrak{A} < b) = (\mathcal{B}U, U)$$

(where again for simplicity we suppose the function U to be normalized).

Thus if we know how to assign a self-adjoint operator to the quantity $\mathfrak{A}$ and also how to assign operators to each $\mathfrak{B}$, then (4) makes it possible to find the complete distribution law of $\mathfrak{A}$.

Furthermore, we note that because of the self-adjoint property of the operator $\mathcal{A}$, we always have

$$(\mathcal{A}U, U) = (U, \mathcal{A}U).$$

On the other hand, because of the general property of scalar products (§1, property 3),

$$(U, \mathcal{A}U) = (\mathcal{A}U, U)^*.$$

Thus we always have

$$(\mathfrak{A}U, U) = (\mathfrak{A}U, U)^*,$$

i.e., the mathematical expectation of a physical quantity is always real. This result shows clearly the meaning of the requirement that all operators which are assigned to physical quantities be self-adjoint. We shall find below even more decisive arguments to support this requirement.

§3. Possible values of physical quantities

Suppose that a physical quantity $\mathfrak{A}$ has a linear self-adjoint operator $\mathcal{A}$ assigned to it. In the preceding section we saw how, knowing this operator $\mathcal{A}$ and the operators corresponding to various functions of the quantity $\mathfrak{A}$, one can construct the distribution law of this quantity in any state U of the system. Sometimes this law can be such that the quantity $\mathfrak{A}$ for the state U takes on only one possible value α (with probability 1). In that case we say that $\mathfrak{A}$ has a *definite* value in the state U. Then

$$\mathsf{E}_U\mathfrak{A} = \alpha.$$

We now consider the conditions under which this situation can occur. In order that a random variable have a unique possible value, it is necessary and sufficient that its dispersion be zero. The dispersion of the quantity $\mathfrak{A}$ in the state U is

$$\mathsf{D}_U\mathfrak{A} = \mathsf{E}_U(\mathfrak{A} - \alpha)^2.$$

On the basis of the principles explained in the preceding section, we naturally conclude that the quantity $(\mathfrak{A} - \alpha)^2$ has a (linear self-adjoint) operator $(\mathcal{A} - \alpha)^2$ corresponding to it. Hence if the function U is normalized,

$$\mathsf{D}_U\mathfrak{A} = ([\mathcal{A} - \alpha]^2U, U).$$

In virtue of the self-adjoint property of $\mathcal{A} - \alpha$ this gives

$$\begin{aligned}\mathsf{D}_U\mathfrak{A} &= ([\mathcal{A} - \alpha]U, [\mathcal{A} - \alpha]U)\\ &= \int ([\mathcal{A} - \alpha]U)([\mathcal{A} - \alpha]U)^*\, dq_1 \cdots dq_s\\ &= \int |\,[\mathcal{A} - \alpha]U\,|^2\, dq_1 \cdots dq_s.\end{aligned}$$

If $\mathsf{D}_U\mathfrak{A} = 0$, then because of the non-negative character of the integrand, we must have

$$|\,[\mathcal{A} - \alpha]U\,|^2 = 0$$

identically in $q_1, \cdots, q_s$. Consequently,

$$[\mathcal{A} - \alpha]U = 0, \qquad \mathcal{A}U = \alpha U \tag{5}$$

for arbitrary $q_1, \cdots, q_s$. Accordingly, (5) is a necessary condition that $\mathfrak{A}$ have a definite value α in the state U. This is also a sufficient condition, as can be easily seen by carrying out all the calculations in the reverse order.

The relation (5) shows that the action of $\mathcal{A}$ on U leads to the multiplication of U by a real constant α. In the theory of operators, a function U (not identically zero) related to an operator $\mathcal{A}$ by an equation of the form (5) is called an *eigenfunction* of the operator belonging to the *eigenvalue* α. The numbers which can occur as definite values of the quantity $\mathfrak{A}$ are the *eigenvalues of the operator* $\mathcal{A}$ and the state in which the quantity $\mathfrak{A}$ has a definite value α is described *by one of the eigenfunctions* U *of the operator* $\mathcal{A}$ *belonging to the eigenvalue* α.

If we consider (5) as an equation defining U, then in typical quantum-mechanical cases this will be a partial differential equation, since the operator $\mathcal{A}$ is formed from the operators $\mathcal{Q}_k$, $\mathcal{P}_k$ in about the same way as the quantity $\mathfrak{A}$ is formed from the variables q_k, p_k $(k = 1, \cdots, s)$. The desired solutions U of this equation must be uniquely determined in the entire space of the variables $q_1, \cdots, q_s$, and the integral

$$\int |U|^2 \, dq_1 \cdots dq_s$$

must exist when taken over the entire space. Moreover, these functions must satisfy certain other general requirements: They must be continuous and sometimes they must possess partial derivatives to some order. Despite the seeming broadness of these requirements, in general it turns out that solutions satisfying them do not exist for all values of the parameter α. Those values of α for which such solutions exist are eigenvalues of the operator $\mathcal{A}$. The set of eigenvalues of $\mathcal{A}$ is called the *spectrum* of $\mathcal{A}$. The spectrum of an operator can be either *discrete* (i.e., consisting of single isolated numbers) or *continuous* (i.e., comprising a whole segment or even the whole continuum of real numbers). Cases of *combined* spectra are also possible, i.e., spectra which in some places are continuous and in other places discrete. For the purposes of this book, we need only consider the case in which the (discrete) spectrum of an operator consists of a sequence of numbers approaching infinity:

$$\alpha_1 < \alpha_2 < \cdots < \alpha_n \cdots \qquad (\lim_{n\to\infty} \alpha_n = \infty). \tag{6}$$

Suppose that in a certain measurement of the quantity $\mathfrak{A}$ we have obtained the value α. According to quantum mechanics, immediately after this measurement the system will be in a state such that the quantity $\mathfrak{A}$ has α as its definite value. Hence a second measurement of $\mathfrak{A}$ on the same system, provided it follows immediately after the first, will necessarily have for its result the value α. Therefore, any number α which can ever appear as the result of a measurement of the quantity $\mathfrak{A}$ (in any state whatever) must also be (for some particular state) a definite value of this quantity and consequently must belong to the spectrum of the operator $\mathfrak{a}$. This spectrum accordingly contains all numbers which can occur as results of a measurement of $\mathfrak{A}$ with the system in an arbitrary state. In the case of the discrete spectrum (6) which we are considering, the discreteness is nowhere specially postulated, but appears as a consequence of very general requirements imposed on the eigenfunctions. This is similar to the theory of an oscillating string in which the discreteness of the sequence of possible modes of vibration follows from the boundary conditions imposed. The determination of all possible values of a physical quantity by constructing the spectrum of its corresponding operator is commonly called the "quantization" of this quantity.

We must now consider certain basic properties of the eigenfunctions of a linear self-adjoint operator $\mathfrak{a}$. We shall confine ourselves to the case in which the spectrum has the form (6) and shall first examine the set of eigenfunctions belonging to a single eigenvalue α_k. Because of the linear property of the operator $\mathfrak{a}$, if U_{k1} and U_{k2} are two different eigenfunctions belonging to the eigenvalue α_k, then $\lambda_1 U_{k1} + \lambda_2 U_{k2}$, where λ_1 and λ_2 are any complex numbers which are not both zero, will also be an eigenfunction belonging to the same eigenvalue α_k. This means that the set of eigenfunctions that we are considering forms a *linear manifold* $\mathfrak{L}_k$. In the cases to be considered in this book, the manifold $\mathfrak{L}_k$ always has a finite number of dimensions m. This means that the manifold $\mathfrak{L}_k$ contains m linearly independent functions $U_{k1}, U_{k2}, \cdots, U_{km}$, but any $m + 1$ functions of this manifold are linearly dependent. Consequently, any function U of the manifold $\mathfrak{L}_k$ can be expressed in a unique way in the form

$$U = \sum_{j=1}^{m} \lambda_j U_{kj},$$

where the λ_j are complex numbers. The functions U_{kj} ($j = 1, 2, \cdots, m$) form a linear basis of the manifold $\mathfrak{L}_k$. This basis can be chosen in various ways. In particular, well-known processes of orthogonalization always make possible the construction of a basis consisting of mutually orthogonal functions U_{kj} [the functions U_1 and U_2 are mutually orthogonal, if $(U_1, U_2) = 0$].

The basis U_{kj} ($j = 1, 2, \cdots, m$) is called *normalized* if $(U_{kj}, U_{kj}) = 1$ ($1 \leq j \leq m$), i.e., if all functions forming this basis are normalized.

Let $V_{k1}, V_{k2}, \cdots, V_{km}$ be any m eigenfunctions of the operator $\mathfrak{A}$, belonging to the same eigenvalue α_k of this operator as the functions U_{kj}. Then,

$$V_{kg} = \sum_{j=1}^{m} \lambda_{gj} U_{kj} \qquad (1 \leq g \leq m), \tag{7}$$

where the λ_{gj} are complex numbers. Hence, for $1 \leq g \leq m$, $1 \leq h \leq m$, we have

$$\begin{aligned}(V_{kg}, V_{kh}) &= \left(\sum_{j=1}^{m} \lambda_{gj} U_{kj}, \sum_{i=1}^{m} \lambda_{hi} U_{ki}\right) \\ &= \sum_{j=1}^{m} \sum_{i=1}^{m} \lambda_{gj} \lambda_{ki}^{*} (U_{kj}, U_{ki}).\end{aligned}$$

Since the basis U_{kj} is orthogonal and normalized,

$$(U_{kj}, U_{ki}) = \delta_{ji},$$

where

$$\delta_{ji} = \begin{cases} 1 & (j = i), \\ 0 & (j \neq i). \end{cases}$$

Therefore, we find

$$(V_{kg}, V_{kh}) = \sum_{j=1}^{m} \sum_{i=1}^{m} \delta_{ji} \lambda_{gj} \lambda_{hi}^{*} = \sum_{j=1}^{m} \lambda_{gj} \lambda_{hj}^{*}.$$

In order that the functions V_{kg}, defined by the relations (7), be orthogonal and normalized, it is necessary and sufficient that

$$\sum_{j=1}^{m} \lambda_{gj} \lambda_{hj}^{*} = \begin{cases} 1 & (g = h), \\ 0 & (g \neq h), \end{cases}$$

or, equivalently, that

$$\sum_{j=1}^{m} \lambda_{gj} \lambda_{hj}^{*} = \delta_{gh} \quad (1 \leq g \leq m, 1 \leq h \leq m). \tag{8}$$

The matrix consisting of the complex numbers λ_{gh} $(1 \leq g \leq m, 1 \leq h \leq m)$ (and also the linear transformation (7) corresponding to this matrix), is called *unitary*, if the conditions (8) hold for it. Since the functions V_{kg} are orthogonal and normalized, they are also linearly independent and consequently form an orthogonal and normalized basis of the manifold $\mathfrak{L}_k$. [Indeed, if, for example, we had

$$V_{k1} = \alpha_2 V_{k2} + \cdots + \alpha_m V_{km},$$

where the α_i are numbers, then we would find

$$(V_{k1}, V_{k1}) = \sum_{l=2}^{m} \alpha_l (V_{kl}, V_{k1}) = 0,$$

which is impossible.] Thus, *if the functions U_{kj} form an orthogonal normalized basis of the manifold $\mathfrak{L}_k$, then all other such bases V_{kg} $(1 \leq g \leq m)$ of this* *manifold are obtained by subjecting the system of functions* U_{kj} $(1 \leq j \leq m)$ *to all possible unitary transformations* (7).

The number of dimensions m of the manifold $\mathfrak{L}_k$ is a very important property of the eigenvalue α_k and is called its *multiplicity* or *degree of degeneracy*. The eigenvalue α_k is *degenerate* if $m > 1$ and *non-degenerate* if $m = 1$. From the physical point of view m gives the number of linearly independent states of the system in which the quantity $\mathfrak{A}$ has the definite value α_k.

Now let U_1 and U_2 be two eigenfunctions of the operator $\mathcal{A}$, belonging to *different* eigenvalues α_{k_1} and α_{k_2}, so that

$$\mathcal{A}U_1 = \alpha_{k_1}U_1, \qquad \mathcal{A}U_2 = \alpha_{k_2}U_2. \tag{9}$$

In virtue of the self-adjoint property of the operator $\mathcal{A}$, we have

$$(\mathcal{A}U_1, U_2) = (U_1, \mathcal{A}U_2).$$

Hence, by (9),

$$(\alpha_{k_1}U_1, U_2) = (U_1, \alpha_{k_2}U_2).$$

Since α_{k_1} and α_{k_2} are real, this last equation and property 2 (§1) of scalar products lead to

$$\alpha_{k_1}(U_1, U_2) = \alpha_{k_2}(U_1, U_2),$$

and the fact that $\alpha_{k_1} \neq \alpha_{k_2}$, to

$$(U_1, U_2) = 0.$$

This shows that *two eigenfunctions of the operator* $\mathcal{A}$, *belonging to different eigenvalues, are always mutually orthogonal.*

If we now choose for each eigenvalue α_k of the operator $\mathcal{A}$ some linear orthogonal basis, then the whole set of eigenfunctions of the operator $\mathcal{A}$ so obtained obviously forms a (denumerably) infinite orthogonal system of elements U of a complex Hilbert space. In the simplest and most important cases, and in particular, for all cases with which we shall be concerned, this orthogonal system is *complete* (or *closed*). This means that every element U orthogonal to all elements of the system we have obtained must be identically zero. The completeness of our system of eigenfunctions is very significant in physics. If we denote by $U_1, U_2, \cdots, U_n, \cdots$ the elements of a complete orthogonal system, enumerated in arbitrary order, then any element U of the complex Hilbert space can be represented in the form of a series

$$\sum_{n=1}^{\infty} c_n U_n$$

which converges to U "in the mean". [This means that setting $\sum_{k=1}^{n} c_k U_k =$

s_n, we have $(U - s_n, U - s_n) \to 0$ as $n \to \infty$.] Here the c_n are complex constants. In virtue of the mutual orthogonality of the functions U_n we easily find

$$c_n = (U, U_n)/(U_n, U_n).$$

If the functions U_n are normalized, then

$$c_n = (U, U_n),$$

and furthermore,

$$(U, U) = \sum_{n=1}^{\infty} c_n(U_n, U) = \sum_{n=1}^{\infty} c_n(U, U_n)^* = \sum_{n=1}^{\infty} | c_n |^2.$$

In particular, if the function U is also normalized, then

$$\sum_{n=1}^{\infty} | c_n |^2 = 1.$$

In the preceding section we saw that the ability to assign to each physical quantity a definite linear self-adjoint operator allows us in principle to determine the statistics (distribution law) of this quantity in any state U. We are now in a position to show concretely how this is done, at least for quantities with a discrete spectrum. Let the operator $\mathcal{A}$, with the discrete spectrum (6), correspond to the quantity $\mathfrak{A}$. Since the only possible values of $\mathfrak{A}$ are the numbers α_k, the statistics of $\mathfrak{A}$ in any state U will be fully determined, if we can find the probability

$$\mathbf{P}_U(\mathfrak{A} = \alpha_k)$$

for each α_k. As in §2, we denote by $\mathfrak{B}$ a quantity equal to 1 if $\mathfrak{A} = \alpha_k$, and equal to zero if $\mathfrak{A} \neq \alpha_k$ (so that $\mathfrak{B}$ is a function of $\mathfrak{A}$). Let $\mathcal{B}$ be the linear self-adjoint operator corresponding to the quantity $\mathfrak{B}$. Then the probability that $\mathfrak{A} = \alpha_k$ is the same as the probability that $\mathfrak{B} = 1$, and is obviously equal to the mathematical expectation of $\mathfrak{B}$.

Thus, we have (assuming the function U to be normalized)

$$\mathbf{P}_U(\mathfrak{A} = \alpha_k) = (\mathcal{B}U, U).$$

Now let $U_1, U_2, \cdots, U_n, \cdots$ be the complete orthogonal system of normalized eigenfunctions of the operator $\mathcal{A}$ and let

$$U = \sum_{n=1}^{\infty} c_n U_n, \qquad c_n = (U, U_n) \qquad (n \geq 1).$$

Then

$$(\mathcal{B}U, U) = \left(\sum_{n=1}^{\infty} c_n \mathcal{B} U_n, \sum_{n=1}^{\infty} c_n U_n\right).$$

Each of the functions U_n is an eigenfunction of the operator $\mathcal{A}$, i.e., the quantity $\mathfrak{A}$ has a definite value in each of the states U_n. From this it follows that the quantity $\mathfrak{B}$, as a function of $\mathfrak{A}$, must also have a definite value in each of the states U_n. This value is equal to 1 in those states for which $\mathfrak{A} = \alpha_k$, and is equal to zero in the other states. Thus, each of the functions U_n is an eigenfunction of the operator $\mathcal{B}$; the corresponding eigenvalue of this operator is 1 if $\mathcal{A}U_n = \alpha_k U_n$, and zero in all other cases. We can therefore write

$$\mathcal{B}U_n = \beta_n U_n,$$

where $\beta_n = 1$ if $\mathcal{A}U_n = \alpha_k U_n$ and $\beta_n = 0$ otherwise. Therefore, we obtain

$$(\mathcal{B}U, U) = (\textstyle\sum_{n=1}^{\infty} c_n \beta_n U_n, \sum_{n=1}^{\infty} c_n U_n).$$

By using the properties of scalar products, the orthogonality of the system of functions U_n, and the normalized character of these functions, we now easily find

$$\mathbf{P}_U(\mathfrak{A} = \alpha_k) = (\mathcal{B}U, U) = \textstyle\sum_{n=1}^{\infty} \beta_n \mid c_n \mid^2$$
$$= \textstyle\sum^{(k)} \mid c_n \mid^2 = \sum^{(k)} \mid (U, U_n) \mid^2,$$

where $\sum^{(k)}$ means that the summation is taken over all values n for which the eigenfunction U_n belongs to the eigenvalue α_k. Applying the theorem on the composition of probabilities to this expression, we find

$$\mathbf{P}_U(a < \mathfrak{A} < b) = \textstyle\sum_{(a,b)} \mid c_n \mid^2 = \sum_{(a,b)} \mid (U, U_n) \mid^2,$$

where the summation is taken over all eigenfunctions U_n of the operator $\mathcal{A}$ belonging to eigenvalues included between a and b. In this way we obtain a definite rule for finding the distribution law of the quantity $\mathfrak{A}$ in the state U: Expand the function U in terms of the complete orthogonal system of normalized eigenfunctions U_n of the operator $\mathcal{A}$. The probability that $a < \mathfrak{A} < b$ is the sum of the squares of the absolute values of the coefficients of this expansion for all those functions U_n that belong to eigenvalues included between a and b.

We now apply the complete orthogonal and normalized system of eigenfunctions of the operator $\mathcal{A}$ to prove one more elementary proposition which we shall require later: *If the operator $\mathcal{A}$ has the spectrum* (6), *then the spectrum of the operator $\mathcal{A}^2$ consists of the numbers*

$$\alpha_1^2, \alpha_2^2, \cdots, \alpha_k^2, \cdots;$$

and the manifold of eigenfunctions of the operator $\mathcal{A}^2$ belonging to the eigenvalue α_k^2 coincides with the manifold of eigenfunctions of the operator $\mathcal{A}$ belonging to the eigenvalue α_k.

Indeed, if U is an eigenfunction of the operator $\mathcal{A}$ belonging to the eigenvalue α_k, then

$$\mathcal{A}^2 U = \mathcal{A}(\mathcal{A}U) = \mathcal{A}(\alpha_k U) = \alpha_k \mathcal{A}U = \alpha_k^2 U,$$

i.e., U is an eigenfunction of the operator $\mathcal{A}^2$ belonging to the eigenvalue α_k^2.

Now suppose that, conversely, U is an eigenfunction of the operator $\mathcal{A}^2$ belonging to some eigenvalue β of this operator.

We expand the function U in terms of the complete orthogonal system of eigenfunctions of the operator $\mathcal{A}$ (which, as we have just ascertained, are at the same time eigenfunctions of the operator $\mathcal{A}^2$):

$$U = \sum_{k=1}^{\infty} c_k U_k , \tag{10}$$

where

$$c_k = (U, U_k) \qquad (k = 1, 2, \cdots).$$

Since U and U_k are eigenfunctions of the same operator $\mathcal{A}^2$, $c_k \neq 0$ only for $\beta = \alpha_k^2$. This shows that an eigenvalue β of the operator $\mathcal{A}^2$ must necessarily coincide with one of the numbers α_k^2, i.e., that the spectrum of the operator $\mathcal{A}^2$ actually gives the set of numbers α_k^2. Now let $\beta = \alpha_k^2$ and let U_{k1}, $U_{k2}, \cdots, U_{kn}$ be the eigenfunctions of the operator $\mathcal{A}$ from the orthogonal system we have chosen, belonging to the eigenvalue α_k. Then the only non-vanishing coefficients in (10) are those of the functions U_{ki} $(1 \leq i \leq m)$. Hence,

$$U = \sum_{i=1}^{m} c_{ki} U_{ki} ,$$

and consequently,

$$\mathcal{A}U = \alpha_k U,$$

i.e., U is an eigenfunction of the operator $\mathcal{A}$, and our proposition is proved.

§4. Evolution of the state of a system in time

In classical mechanics the state of a system is described by the values of all $2s$ Hamiltonian variables $q_1, \cdots, q_s, p_1, \cdots, p_s$. In order to know how this state changes with time, it is necessary to give all $2s$ of these variables as functions of time. The fact that the values of the Hamiltonian variables at any initial time t_0 uniquely determine their values at any other (preceding or following) time t is an expression of the principle of causality in classical mechanics.

In quantum mechanics the state of a system is described by a certain function $U(q_1, q_2, \cdots, q_s)$ of the generalized coordinates. The change of the state with time will be known if the dependence of the function U on the time is given, i.e., if U is given as a function of the $s + 1$ variables $q_1, \cdots, q_s, t$:

$$U = U(q_1, \cdots, q_s, t).$$

In quantum mechanics an expression of the principle of causality is the requirement that the form of the function U at any initial time t_0 uniquely determine its form at any other (preceding or following) time t. Since, in quantum mechanics, the function U determines the statistics (distribution law) of physical quantities and not their values, this requirement has the physical meaning that the statistics of physical quantities, at the initial time t_0, uniquely determine their statistics at any other time t.

In classical mechanics the law of motion (i.e., the expression of the Hamiltonian variables as functions of time) is found by solving the "equations of motion". In the case of Hamiltonian variables these equations assume the well-known "canonical" form, which is remarkably simple and symmetric. The fundamental role in these equations is played by the so-called Hamiltonian function

$$H(q_1, \cdots, q_s, p_1, \cdots, p_s).$$

When the forces acting on the system are independent of time, the expression for the Hamiltonian function is the same as the expression for the total energy of the system

$$H = T + V,$$

where T is the kinetic energy, and V the potential energy of the system.

In quantum mechanics the Hamiltonian function is assigned a linear self-adjoint operator $\mathcal{H}$, the operator of total energy, also called the *Hamiltonian*. This operator is just as important for the determination of the evolution of the state of a system in time as the Hamiltonian function is in classical mechanics.

In the simplest cases the expression for the operator $\mathcal{H}$ can be chosen correctly by direct analogy with the expression for the Hamiltonian function in classical mechanics. If our system is an elementary particle, then the simplest procedure is to choose as the Hamiltonian variables its three Cartesian coordinates x, y, z and the corresponding components of momentum p_x, p_y, p_z. In this case the kinetic energy will be

$$T = (p_x^2 + p_y^2 + p_z^2)/2m,$$

where m is the mass of the particle, and the potential energy $V(x, y, z)$ will depend only on the coordinates of the particle. Hence the expression for the Hamiltonian function is

$$H = (1/2m)(p_x^2 + p_y^2 + p_z^2) + V(x, y, z).$$

To construct the operator $\mathcal{H}$, we recall that the operators corresponding to the quantities p_x, p_y, p_z are $-i\hbar\, \partial/\partial x$, $-i\hbar\, \partial/\partial y$, $-i\hbar\, \partial/\partial z$. According to the algebra of operators adopted in §2, the operators corresponding to the squares of these quantities must be

$$-\hbar^2\, \partial^2/\partial x^2, \qquad -\hbar^2\, \partial^2/\partial y^2, \qquad -\hbar^2\, \partial^2/\partial z^2.$$

Thus, the operator corresponding to the kinetic energy T will be

$$-(\hbar^2/2m)(\partial^2/\partial x^2 + \partial^2/\partial y^2 + \partial^2/\partial z^2) = -(\hbar^2/2m)\nabla^2,$$

where $\nabla^2 = \partial^2/\partial x^2 + \partial^2/\partial y^2 + \partial^2/\partial z^2$ is the so-called Laplacean operator. Since the operator corresponding to the potential energy $V(x, y, z)$ is simply the operator $\mathcal{V}$ of multiplication by $V(x, y, z)$, we naturally take as the Hamiltonian operator

$$\mathcal{H} = -(\hbar^2/2m)\nabla^2 + \mathcal{V}. \tag{11}$$

The operator $\mathcal{H}$ is constructed just as simply in the somewhat more complicated case of a system consisting of several such structureless physical particles, if it is assumed that interaction forces between the particles are absent. In this case the total energy of the system is simply the sum of the total energies of the particles composing the system. We naturally assume that the Hamiltonian of this system is equal to the sum of the Hamiltonians of the particles composing the system where the latter are constructed according to equation (11).

Such an immediately obvious determination of the operator $\mathcal{H}$ is possible only in the simplest cases. Even if the system consists of a single structureless particle, the procedure is this simple only when rectangular coordinates are chosen as the Hamiltonian variables. In other coordinate systems the kinetic energy is a quadratic form in the generalized momenta, in which the coefficients depend on the generalized coordinates. In this case the proper choice for the corresponding operator is not so obvious.

We now turn to a discussion of the laws of quantum mechanics which determine the change of state of a system with time. For brevity, we denote the function $U(q_1, \cdots, q_s, t)$, which describes the state of a system at time t, by $U(q, t)$. If, as we require, the specification of the function $U(q, t_0)$ uniquely determines the function $U(q, t)$ for all t, then in particular the function $\partial U/\partial t$ is uniquely determined at $t = t_0$. (We assume, of course, that this derivative exists.) Since t_0 was chosen arbitrarily, we can therefore say that for any t some function $\partial U/\partial t$ uniquely corresponds to each function U. We have agreed to call such a correspondence an operator. Thus the function $\partial U/\partial t$ is the result of the action of a certain operator on the function U. In general, this operator can be different for different times, so that it will be convenient to denote it by $\mathcal{A}_t$. Hence,

$$\partial U(q, t)/\partial t = \mathcal{A}_t U(q, t). \tag{12}$$

We still have to determine the operator $\mathcal{A}_t$. By a heuristic method, which involves the analysis of a number of simple special examples and the generalization of the results so obtained (and includes the application of a number of theoretical considerations), it has been found possible to establish the universal form of this operator. It turns out that one must set

$$\mathfrak{A}_t = -i\mathcal{H}/\hbar,$$

where $\hbar$ is Planck's constant divided by 2π and $\mathcal{H}$ is the Hamiltonian of the system. This rule is always valid. In particular, it remains true when variable forces act on the system, i.e., when the operator $\mathcal{H}$ depends on the time and is, in general, no longer the operator of total energy. However, in all that follows we shall suppose that the operator $\mathcal{H}$ is independent of the time, and, consequently, that it corresponds to the physical quantity which we call the total energy of the system. Equation (12) then takes the form

$$i\hbar\, \partial U(q, t)/\partial t = \mathcal{H}U(q, t),$$

or, more briefly,

$$i\hbar\, \partial U/\partial t = \mathcal{H}U. \tag{13}$$

This is the most important equation in quantum mechanics, and is analogous to the system of equations of motion of classical mechanics. It is commonly called the Schrödinger equation. It defines the function $U(q_1, \cdots, q_s, t)$ as the solution of a certain partial differential equation. We shall see that this is always an equation of first order in t, but of the second order in the generalized coordinates q_k, since the kinetic energy operator which forms part of the operator $\mathcal{H}$ contains the operators $\partial^2/\partial q_k^2$.

As our first application of Schrödinger's equation we prove an important auxiliary theorem which will be needed later. We call the quantity

$$(U, U) = \int |U|^2\, dq_1 \cdots dq_s$$

the *norm* of the function U so that in particular, for normalized functions U, the norm is equal to unity. We now show that the norm of any solution to Schrödinger's equation remains constant in time (i.e., the functional (U, U) is an integral of the Schrödinger equation). In particular, if the function U which describes the state of a system is normalized $[(U, U) = 1]$ at t_0, then it will also be normalized at any other (preceding or following) time t.

To prove this, we start with equation (13), where $U = U(q_1, q_2, \cdots, q_s, t)$. Forming the complex conjugate quantities yields

$$-i\hbar\, \partial U^*/\partial t = (\mathcal{H}U)^*. \tag{14}$$

We multiply equations (13) and (14) by U^* and U, respectively, and subtract the second equation from the first. We find

$$i\hbar[U^*(\partial U/\partial t) + U(\partial U^*/\partial t)] = i\hbar\, \partial \mid U \mid^2/\partial t = (\mathcal{H}U)U^* - U(\mathcal{H}U)^*.$$

Integrating this equation over the entire configuration space, we obtain

$$i\hbar[d(U, U)/dt] = (\mathcal{H}U, U) - (U, \mathcal{H}U).$$

The right side of this equation is equal to zero, because of the self-adjoint character of the operator $\mathcal{H}$. Therefore,

$$d(U, U)/dt = 0,$$

which was to be proved.

§5. Stationary states. The law of conservation of energy

In the cases of interest to us, the spectrum of the total energy operator $\mathcal{H}$ of the system will always consist of a sequence of numbers of the form (6) of §3. Therefore, we can select a linear orthogonal basis

$$U_1, U_2, \cdots, U_n, \cdots$$

of normalized eigenfunctions of the operator $\mathcal{H}$. Let the function U_n belong to the eigenvalue E_n of the operator $\mathcal{H}$, so that

$$\mathcal{H}U_n = E_nU_n \qquad (n = 1, 2, \cdots). \tag{15}$$

(In general, because of the possibility of degeneracy, the numbers E_n may contain sets of successive numbers that are equal to each other.)

A solution $U(q, t)$ of Schrödinger's equation can then be expanded in terms of the functions U_n. The coefficients of this expansion will, of course, vary with t, so that we can write

$$U(q, t) = \sum_{n=1}^{\infty} a_n(t)U_n(q),$$

where $a_n(t)$ are complex functions of t. Substituting this expression into Schrödinger's equation (13) and assuming, as we have agreed, that the operator $\mathcal{H}$ is independent of time, we find in virtue of equation (15)

$$i\hbar \sum_{n=1}^{\infty} (da_n/dt)U_n(q) = \sum_{n=1}^{\infty} a_n(t)\mathcal{H}U_n(q) = \sum_{n=1}^{\infty} E_na_n(t)U_n(q).$$

Because of the uniqueness of expansions in terms of the functions $U_n(q)$ we can equate the coefficients of $U_n(q)$ on the left and right sides:

$$i\hbar\, da_n/dt = E_na_n(t) \qquad (n = 1, 2, \cdots).$$

From this we readily find

$$a_n(t) = c_ne^{-iE_nt/\hbar}$$

Consequently,

$$U(q, t) = \sum_{n=1}^{\infty} c_n e^{-iE_n t/\hbar} U_n(q), \tag{16}$$

where the c_n are complex constants. Since we have taken the functions $U_n(q)$ to be normalized, it follows that

$$(U, U) = \sum_{n=1}^{\infty} | c_n |^2.$$

Accordingly, if we wish the function $U(q, t)$ to be normalized, we must have

$$\sum_{n=1}^{\infty} | c_n^2 | = 1. \tag{17}$$

Thus, the general normalized solution of Schrödinger's equation can be represented in the form (16), where the complex constants c_n must satisfy (17) for the solution to be normalized. Conversely, if the c_n are arbitrary complex numbers satisfying (17), then (16) gives us a normalized solution of Schrödinger's equation.

In particular, the functions

$$e^{-iE_n t/\hbar} U_n(q) \qquad (n = 1, 2, \cdots) \tag{18}$$

are normalized solutions of Schrödinger's equation. The states of the system described by these functions are commonly called *stationary* states. This name is justified by the fact that if the state of a system is described by a function of the form (18), then the states of this system at the times t_1 and t_2 are described by functions that differ from each other only by a constant factor, and, therefore, the statistics of any physical quantity will be precisely the same at these two times. In other words, in a stationary state the statistics of physical quantities are independent of time. Since these statistics are all that we can obtain from a knowledge of the state of a system in quantum mechanics, we must conclude that when the evolution of a system is described by a function of the form (18), the physical state of this system remains unchanged in the course of time.

Functions of the form (18) are eigenfunctions of the total energy operator $\mathcal{H}$ and, consequently, describe states of the system in which its total energy has a definite value. In this way we arrive at the important conclusion that, if the total energy of a system at a certain instant of time has a definite value (for example, if at that instant a measurement of the total energy has just been made), then the physical state of the system in its further evolution will remain unchanged. This conclusion is particularly interesting since there is no analogous result in classical mechanics. Indeed, in classical mechanics the total energy always has a definite value and does not change with time (if the external forces remain constant). The Hamiltonian variables, however, which evolve according to the equations of motion, constantly change their values; and since the state of the system in classical mechanics is defined by a set of these values (or, equivalently, by a point in its phase space), this state is subject to continuous change. The various "ergodic" theorems or hypotheses even try to establish that this change has an extremely broad character, i.e., that the state of the system in the course of time comes arbitrarily close to any state consistent with the given value of the total energy. We now see that in quantum mechanics similar ergodic postulates or theorems are impossible, at least in states in which the total energy has a definite value. However, statistical thermodynamics is primarily concerned with just such states.

We must now determine the physical quantities in quantum mechanics which are integrals of the motion. In classical mechanics a function of the Hamiltonian variables whose value, in virtue of the equations of motion, remains constant in time is called an integral of the motion. In quantum mechanics Schrödinger's equation serves as the equation of motion. The function $U(q, t)$ determined by this equation does not permit us to find the values of physical quantities, but only their distribution laws (i.e., their statistics). Therefore, Schrödinger's equation can have as a consequence the constancy in time of at most the distribution law of a physical quantity, but not of its value which in general is not determined. Hence in quantum mechanics we must call any physical quantity, whose distribution law is time independent, an integral of the motion.

The mathematical apparatus of quantum mechanics yields a characterization for integrals of the motion which is remarkably simple. It is expressed by the following proposition:

THEOREM. *In order for a physical quantity $\mathfrak{A}$ to be an integral of the motion, it is necessary and sufficient that the corresponding operator $\mathcal{A}$ commute with the total energy operator $\mathcal{H}$ of the system, i.e., that $\mathcal{A}\mathcal{H} = \mathcal{H}\mathcal{A}$.*

We prove only the sufficiency of this criterion, since we shall not require its necessity. For this purpose, we first prove the following very general auxiliary proposition:

LEMMA. *Let $\mathcal{A}$ and $\mathcal{B}$ be two linear self-adjoint operators with discrete spectra, and let $\mathcal{A}\mathcal{B} = \mathcal{B}\mathcal{A}$. Then there exists a complete orthogonal system of eigenfunctions of the operator $\mathcal{A}$, such that all of the functions of this system are simultaneously eigenfunctions of the operator $\mathcal{B}$.*

Proof of the Lemma. We begin with an arbitrary complete orthogonal system of eigenfunctions of the operator $\mathcal{A}$. Let α be any eigenvalue of the operator $\mathcal{A}$ and let U be any eigenfunction of the operator belonging to this eigenvalue, so that

$$\mathcal{A}U = \alpha U.$$

If $\mathcal{A}\mathcal{B} = \mathcal{B}\mathcal{A}$, then

$$\mathcal{A}\mathcal{B}U = \mathcal{B}\mathcal{A}U = \mathcal{B}(\alpha U) = \alpha\mathcal{B}U,$$

i.e., the function $\mathcal{B}U$ *is also an eigenfunction of the operator* $\mathcal{A}$ *belonging to the same eigenvalue* α.

Now suppose the eigenvalue α of the operator $\mathcal{A}$ has multiplicity (degree of degeneracy) m, and let the eigenfunctions belonging to this eigenvalue in our chosen complete orthogonal system be $U_1, U_2, \cdots, U_m$. We have just shown that the functions $\mathcal{B}U_k$ $(k = 1, 2, \cdots, m)$ are also eigenfunctions of the operator $\mathcal{A}$ belonging to the eigenvalue α. Consequently, we must have

$$\mathcal{B}U_k = \sum_{i=1}^{m} a_{ik} U_i \qquad (k = 1, 2, \cdots, m), \tag{19}$$

where the a_{ik} are complex numbers.

Now let

$$V = \sum_{k=1}^{m} c_k U_k$$

be a linear combination of the functions U_k with arbitrary complex coefficients c_k, so that V is also an eigenfunction of the operator $\mathcal{A}$ belonging to the eigenvalue α. We shall try to choose these coefficients so that the function V will also be an eigenfunction of the operator $\mathcal{B}$. For this to be true, we must have

$$\mathcal{B} \sum_{k=1}^{m} c_k U_k = \beta \sum_{k=1}^{m} c_k U_k , \tag{20}$$

where β is a complex number. But because of (19) and the linearity of the operator $\mathcal{B}$,

$$\mathcal{B} \sum_{k=1}^{m} c_k U_k = \sum_{k=1}^{m} c_k \mathcal{B} U_k = \sum_{k=1}^{m} c_k \sum_{i=1}^{m} a_{ik} U_i ;$$

or, by changing the designations of the summation indices,

$$\mathcal{B} \sum_{k=1}^{m} c_k U_k = \sum_{i=1}^{m} c_i \sum_{k=1}^{m} a_{ki} U_k = \sum_{k=1}^{m} \left(\sum_{i=1}^{m} c_i a_{ki}\right) U_k .$$

Therefore, (20) gives

$$\sum_{k=1}^{m} \left[\sum_{i=1}^{m} c_i a_{ki} - \beta c_k\right] U_k = 0,$$

from which, by using the linear independence of the functions $U_1, \cdots, U_m$, we have

$$\sum_{i=1}^{m} c_i a_{ki} - \beta c_k = 0 \qquad (k = 1, 2, \cdots, m).$$

Writing as before,

$$\delta_{ik} = \begin{cases} 1 & (i = k), \\ 0 & (i \neq k), \end{cases}$$

we can rewrite the above system of equations in the form

$$\sum_{i=1}^{m} (a_{ki} - \beta \delta_{ki}) c_i = 0 \qquad (k = 1, 2, \cdots, m).$$

Accordingly, we obtain a system of homogeneous linear equations for the determination of the coefficients c_i. In order that this system have a non-trivial solution, it is necessary that its determinant be zero. This, obviously, gives an equation of degree m in the parameter β. Let $\beta_1, \beta_2, \cdots, \beta_m$ be the roots of this equation. To the root β_k there then corresponds a system of coefficients c_{ik} $(1 \leq i \leq m)$, for which

$$V_k = \sum_{i=1}^{m} c_{ik} U_i \qquad (k = 1, 2, \cdots, m)$$

is an eigenfunction of the operator $\mathfrak{B}$, belonging to the eigenvalue β_k. (For the sake of simplicity we shall not consider the case in which the numbers β_k are not all distinct. All of our conclusions remain valid in this case and merely require a little more discussion.) At the same time V_k is also an eigenfunction of the operator $\mathfrak{A}$, belonging to the eigenvalue α. The system of functions $V_1, V_2, \cdots, V_m$ can be used to replace our originally chosen system $U_1, U_2, \cdots, U_m$ as the linear basis of the manifold of eigenfunctions of the operator $\mathfrak{A}$, belonging to the eigenvalue α (the orthogonality of this basis follows from the fact that the V_k are eigenfunctions of the operator $\mathfrak{B}$, belonging to different eigenvalues β_k of $\mathfrak{B}$). After making such a replacement for all eigenvalues α of the operator $\mathfrak{A}$, we obtain, by means of the functions V_k, a complete orthogonal system of eigenfunctions of the operator $\mathfrak{A}$. This verifies the lemma stated above.

Proof of the Theorem. Now let $U_1, U_2, \cdots$ be a complete orthogonal system of functions all of whose members are simultaneously eigenfunctions of the operators $\mathfrak{A}$ and $\mathfrak{H}$. (This system exists in virtue of the lemma proved above.) Expanding the solution $U(q, t)$ of Schrödinger's equation in terms of this orthogonal system of functions we obtain

$$U(q, t) = \sum_{k=1}^{\infty} a_k(t) U_k(q),$$

where $a_k(t)$ are complex numbers which may depend on t. Since the $U_k(q)$ are eigenfunctions of the operator $\mathfrak{A}$, the general result of §3 shows that the probability that the quantity $\mathfrak{A}$ assume a particular one of its possible values is expressed uniquely by means of the quantities $|a_k(t)|^2$. On the other hand, the functions $U_k(q)$ are also simultaneously eigenfunctions of the operator $\mathfrak{H}$, so that, as we saw at the beginning of the present section,

$$a_k(t) = c_k e^{-iE_k t/\hbar},$$

where the c_k are complex constants.

From the above it follows that $|a_k(t)|^2 = |c_k|^2$ and, consequently, that the numbers $|a_k(t)|^2$, and also the probabilities mentioned above, do not change with time. This obviously means that the quantity $\mathfrak{A}$ is an integral of the motion, and the proof of the theorem is thus complete.

Since the operator $\mathfrak{H}$ commutes with itself, it follows as a special result of the above theorem that for conservative systems (i.e., systems in which the total energy does not depend explicitly on the time) the Hamiltonian function $H(q_1, \cdots, q_s, p_1, \cdots, p_s)$ is always an integral of the motion. Thus for a conservative system the distribution law of the total energy is time independent.

We must regard this result as the expression of *the law of conservation of energy* in quantum mechanics.

Chapter III

GENERAL PRINCIPLES OF QUANTUM STATISTICS

§1. Basic concepts of statistical methods in physics

In the following, we shall call physical theories *phenomenological* if they are developed independently of the atomistic model of the structure of matter. In phenomenological theories, the state of a system is specified by the values of a small number of variables which characterize that system. Thus, the state of a given mass of gas, not under the influence of any external force field, is completely specified by the values of its volume and temperature, so that every other variable characterizing its state (e.g., the pressure of the gas) is a function of these two basic characteristics. From the point of view of a statistical theory, the state of a gas is not uniquely defined by giving its volume and temperature, because there are countless different combinations of positions and velocities of the particles of the gas consistent with given values of the volume and temperature. The state of a gas is uniquely defined in a classical statistical theory only when the positions and velocities of all of its particles are given. Thus each distinct state in the sense of a phenomenological theory comprises a countless set of states which are distinct in the sense of a statistical theory. This fact should cause no essential difficulties if every physical quantity which is of interest in the phenomenological theory (and hence a function of the volume and temperature of the gas) has, in the sense of the statistical theory, the same value in all those different states consistent with the given volume and temperature. However, this is not so; the statistically defined pressure of a gas, for example, can be very different for different combinations of positions and velocities of the particles of the gas, all of which are consistent with the given volume and temperature.

The situation is precisely the same in the general case. From the point of view of a phenomenological theory the state of a physical system is completely described by the values of a small number of physical quantities $\mathfrak{A}^{(1)}, \mathfrak{A}^{(2)}, \cdots, \mathfrak{A}^{(k)}$. Every other quantity $\mathfrak{B}$, which characterizes the state of the system, is a function of the quantities $\mathfrak{A}^{(1)}, \mathfrak{A}^{(2)}, \cdots, \mathfrak{A}^{(k)}$. In a statistical theory, one usually has not one, but countless states of the system consistent with given values of the quantities $\mathfrak{A}^{(1)}, \mathfrak{A}^{(2)}, \cdots,$ $\mathfrak{A}^{(k)}$. In these states, as a rule, the quantity $\mathfrak{B}$ will assume different values. (In quantum physics, besides this problem there is another specific difficulty: In a statistically defined state, a given quantity does not, in general, have a definite value, but has only a certain probability distribution. We will not dwell on this difficulty at the moment.) Now, if the statistical theory is to provide a foundation for the phenomenological theory, then it must be capable of predicting the values which $\mathfrak{B}$ can assume for fixed values of the quantities $\mathfrak{A}^{(1)}, \mathfrak{A}^{(2)}, \cdots, \mathfrak{A}^{(k)}$ and of explaining why experiments (in agreement with the phenomenological theory) invariably give the same value for $\mathfrak{B}$.

We shall examine this situation in a little more detail. If the values of the quantities $\mathfrak{A}^{(1)}, \mathfrak{A}^{(2)}, \cdots, \mathfrak{A}^{(k)}$ are known, then from the point of view of the phenomenological theory the system is in a completely defined state U. (This state should not be confused with the quantum-mechanical state of a system for which the symbol U was also used.) From the point of view of the statistical theory, U represents not one state but a whole family of states $U_1, U_2, \cdots$. (This family, in general, has the power of the continuum, so that it is impossible to enumerate its members. We avoid the general case here only for simplicity of writing; the loss of generality is of no importance to us at this point.) The quantity $\mathfrak{B}$ which in a phenomenological theory has a completely defined value in the state U, takes on, in general, different values B_i in the different states U_i ($i = 1, 2, \cdots$). How can we reconcile these facts?

Suppose that we measure the quantity $\mathfrak{B}$ many times in succession, in such a way that from the phenomenological point of view the system stays in state U. (It is immaterial whether we experiment on one system at different times or on a whole set of identical systems at the same time.) From the statistical viewpoint, we shall be concerned in these experiments with different states $U_1, U_2, \cdots$ of the system, and so the results $B_1, B_2, \cdots$ will not all be the same. To the question "What is the value of the quantity $\mathfrak{B}$ in state U?" the statistical theory can only respond with an arithmetic average of the results obtained. If we make n measurements and if the system is found to be in state U_1 exactly n_1 times, state U_2 exactly n_2 times, and so on, then we answer that in state U

$$\mathfrak{B} = (n_1 B_1 + n_2 B_2 + \cdots)/n = \lambda_1 B_1 + \lambda_2 B_2 + \cdots,$$

where the λ_i's are the relative frequencies of the states U_i in this set of experiments.

The numbers $B_1, B_2, \cdots$ are, of course, given by the theory. But the numbers λ_i — the relative frequencies of the different states U_i which are members of the family U — cannot be determined by any theory. They depend on the conditions of the experiment and change completely from one set of experiments to another.

This state of affairs can be described further in the following way: From the point of view of the statistical theory, U is a family of states U_i, with corresponding values B_i of the quantity $\mathfrak{B}$. Hence, we can merely say that $\mathfrak{B}$ has a certain *mean* value

$$<\mathfrak{B}> = \sum \lambda_i B_i \qquad (\lambda_i \geq 0, \sum_i \lambda_i = 1)$$

in the state U. The choice of the statistical "weights" λ_i amounts to the choice of a *principle of averaging*. In general, it is clear that different methods of averaging will lead to different mean values of $\mathfrak{B}$. In order for the value of $\mathfrak{B}$ predicted by the theory to agree with the arithmetic average of the results of a given set of experiments, the statistical weight λ_i should agree with the relative number ("quota") of those experiments in which the system is found in state U_i (and hence for which $\mathfrak{B}$ has the value B_i).

From what has been said, it is clear that this situation is really difficult. On the one hand, the phenomenological theory definitely ascribes to the quantity $\mathfrak{B}$ a unique value in state U (i.e., in all the states U_i), which this quantity must therefore assume in each experiment. On the other hand, the statistical theory asserts that in different experiments the quantity $\mathfrak{B}$ will assume different values, and, as an analog to the phenomenological value of $\mathfrak{B}$, it can at best offer a mean value $<\mathfrak{B}>$ of the possible results B_i of the measurements.

The situation is actually worse than this. It turns out that for the calculation of this mean value $<\mathfrak{B}>$, the statistical theory cannot even offer a definite method of averaging (set of statistical weights λ_i), since from its point of view the method of averaging must depend on the conditions of experimentation and can change from one set of experiments to another.

There is only one way to reconcile these contradictions which threaten the scientific theory: to suppose that the quantity $\mathfrak{B}$ assumes identical (or almost identical) values in all (or almost all) of the states U_i which appear as members of the family U. Actually, if almost all of the numbers B_i are very near to one another, then one can expect, for any experimental conditions, that almost all of the results obtained will be nearly the same. Under these conditions, the mean value $<\mathfrak{B}>$ will be independent, within very broad limits, of the principle of averaging we select; thus, it can be a suitable analog to the value given by the phenomenological theory to the quantity $\mathfrak{B}$ in the state U.

In constructing a statistical theory we follow just this path, as no other is possible. First of all, we construct a method which yields a strictly determined procedure for finding the mean value $<\mathfrak{B}>$ of the quantity $\mathfrak{B}$ for given values of $\mathfrak{A}^{(1)}, \mathfrak{A}^{(2)}, \cdots, \mathfrak{A}^{(k)}$. (These $\mathfrak{A}^{(i)}$'s usually include the energy of the system and the so-called "external parameters". Among the latter the most important is the volume occupied by the system.) Then we must demonstrate the *suitability* of this mean value $<\mathfrak{B}>$. The demonstration consists in showing that in the overwhelming majority of states consistent with given values of the quantities $\mathfrak{A}^{(1)}, \mathfrak{A}^{(2)}, \cdots, \mathfrak{A}^{(k)}$, the quantity $\mathfrak{B}$ assumes values very near to $<\mathfrak{B}>$. After this is done, all difficulties disappear. The mean value $<\mathfrak{B}>$ is also that value which $\mathfrak{B}$ must assume, according to the phenomenological theory, whenever the quantities $\mathfrak{A}^{(1)}$, $\mathfrak{A}^{(2)}, \cdots, \mathfrak{A}^{(k)}$ take on this fixed set of values. The statistical theory thus predicts that in an overwhelming majority of experiments the quantity $\mathfrak{B}$ will have a value near $<\mathfrak{B}>$.

However, this solution of the problem, apparently so simple, calls for a great many reservations. First, it is necessary to keep in mind that in all the foregoing the quantity $\mathfrak{B}$ cannot be *any* quantity which takes on a definite value in each state which the statistical theory assigns to the system. Thus, for example, if $\mathfrak{B}$ is the velocity of a definite particle belonging to the system, then none of what was said above is applicable. Although this quantity also has a definite mean value $<\mathfrak{B}>$, one cannot argue for the suitability of this average, because this would mean that for all, or almost all, measurements, the velocity of the chosen particle would maintain an approximately constant value. But in regard to such quantities as the velocity of a discrete particle, everything described above is entirely inapplicable, because, from the point of view of the phenomenological (not statistical) theory, the very concept of a particle does not exist. It is immediately clear that every quantity $\mathfrak{B}$, which can be defined in terms of the phenomenological theory must, in the light of the statistical theory, depend *symmetrically* on the states of all (identical) particles comprising the system. Thus, it makes sense to question the suitability of the mean value only for such quantities, and for them the question receives a favorable answer (at least in the most important and most frequently met cases).

It is also necessary to bear in mind that the method which is selected in the statistical theory for the construction of mean values of physical quantities is a very natural one, but is nevertheless largely arbitrary. Of course, the fact that the mean values obtained by this method are found, post factum, to be suitable, can be used to justify the principle of averaging. (Because of this suitability, clearly, any other method of averaging must lead, as a rule, to the same mean value.) However, this question requires more careful study. When we state that the value of $\mathfrak{B}$ is near $<\mathfrak{B}>$ in *almost all* states of the system, we nevertheless suppose (as we are forced to suppose by considering the known facts) that $\mathfrak{B}$ differs substantially from $<\mathfrak{B}>$ in a certain *small* exceptional set of states. What does the word "small" mean here? How do we estimate the size of this set?

As we saw above, in choosing a principle of averaging we assign "weights" λ_i to the states U_i which are members of the family U. A set of such states is acknowledged to be "small" if the sum of the weights of all states belonging to this set is small. Thus, one can say that every principle of averaging ascribes a definite weight to any set of states U_i. Therefore, when we said above that $\mathfrak{B} \approx <\mathfrak{B}>$ ($\approx$ means "approximately equal to") in *almost all* states U_i, the precise meaning of this statement was the following:

$\mathfrak{B} \approx \langle\mathfrak{B}\rangle$ for all states U_i, with the exception of a certain set M of very small weight. Let us now consider some new method of averaging in which all sets of states receive new weights. If the exceptional set M, whose weight for the old mode of averaging was small, receives a very small weight for the new mode as well, then, as before, the states for which $\mathfrak{B} \approx \langle\mathfrak{B}\rangle$ receive overwhelming weight, and the mean value given by the new method will not be essentially different from $\langle\mathfrak{B}\rangle$.

Thus, from the suitability of the mean value $\langle\mathfrak{B}\rangle$, found from one method of averaging, we can infer that any new method will give approximately the same mean value, *if sets of states whose weights were small in the old method of averaging do not receive large weights in the new method.* If this is not the case, the new method can lead to a mean value significantly different from the old. If this situation is idealized mathematically, we can say that the choice of a principle of averaging is equivalent to the establishment of a measure in the "space" of states U_i; and that, as the condition for the practical equivalence of the mean values of a quantity for two different principles of averaging, we can take the mutual absolute continuity of the two corresponding measures. (A set of measure zero with respect to the first measure must have measure zero with respect to the second measure as well.)

In quantum statistics there is a case, very instructive and having remarkable consequences, in which a new principle of averaging replaced an old one because of certain fundamental considerations. The new principle led to a measure which was not absolutely continuous with respect to the previous one. The new mean values actually proved to be substantially different from the old ones and in many cases gave considerably better agreement with experiment. This situation occurred in the transition from the usual statistical concepts to the so-called "new statistics" — symmetric and antisymmetric. In the next section we shall have occasion to consider this example in complete detail.

Finally, to follow our plan in the domain of quantum physics, it is very important for us to consider the following circumstance: Even for the most accurately defined states of the system, we do not as a rule know the exact values of physical quantities but only their distribution laws (and, in particular, their mathematical expectations). This circumstance, as we shall see, does not prevent our establishing the suitability of mean values for the more important of these quantities, but merely complicates the necessary calculations.

§2. Microcanonical averages

Following the program we have planned, we shall now establish the principle of averaging which will be used throughout the book.

We shall deal only with those states in which the total energy E of our system is precisely fixed ("stationary" states). As we see from Chapter II, the set of such states is described in quantum physics by the set $\mathfrak{M}_E$ of eigenfunctions of the energy operator $\mathcal{H}$ of the system, corresponding to the eigenvalue E of this operator. Let $U_1, U_2, \cdots, U_m$ be a complete orthonormal system of functions of this family. Then any normalized function U of this family can be represented (in a unique manner) in the form

$$U = \sum_{k=1}^{m} \alpha_k U_k, \tag{1}$$

where $\alpha_1, \alpha_2, \cdots, \alpha_m$ are complex numbers satisfying the relation

$$\sum_{k=1}^{m} |\alpha_k|^2 = 1; \tag{2}$$

and, conversely, every function of the form (1), where the numbers α_k are subject to condition (2), is one of the (normalized) eigenfunctions of the operator $\mathcal{H}$, corresponding to the eigenvalue E of this operator. In this way, the family of stationary states of the system having a total energy E is placed in one-to-one correspondence with the complex m-dimensional sphere (2). To establish a measure in this space, we can therefore choose any measure on the sphere (2). The most natural choice, of course, is the following: We assume that

$$\alpha_k = r_k e^{i\varphi_k} \qquad (r_k \geq 0, \ \sum_{k=1}^{m} r_k^2 = 1, \quad 0 \leq \varphi_k \leq 2\pi, \quad 1 \leq k \leq m),$$

and denote by S the real sphere

$$\sum_{k=1}^{m} r_k^2 = 1.$$

Then, in the space of the moduli r_k, we establish the natural Euclidean metric of the real sphere S, and we assume that the phases φ_k are uniformly distributed on the interval $(0, 2\pi)$, independent of one another and of the moduli. Thus, the volume element of the complex sphere (2) for this measure has the form

$$dS\, d\varphi_1 \cdots d\varphi_m,$$

where dS is the "surface element" of the real sphere S. In other words, if M is any measurable set of points on the complex sphere (2), and if $\Psi(\alpha_1, \cdots, \alpha_m)$ is the characteristic function of the set M [i.e., $\Psi(\alpha_1, \cdots, \alpha_m)$ is equal to 1 or 0, depending upon whether or not the point $(\alpha_1, \cdots, \alpha_m)$ belongs to the set M], then the measure of the set M is proportional to the integral

$$\int_S dS \int_0^{2\pi} \cdots \int_0^{2\pi} \Psi(\alpha_1, \alpha_2, \cdots, \alpha_m)\, d\varphi_1\, d\varphi_2 \cdots d\varphi_m.$$

We normalize the measure established in this manner, so that the measure of the entire complex sphere (2) (and, hence, also the measure of every family of stationary states which we consider) is unity. [It is easily seen that the normalizing factor is $(2\pi)^{-m}\Sigma^{-1}$, where $\Sigma = 2\pi^{\frac{1}{2}m}/\Gamma(\frac{1}{2}m)$ is the Euclidean "area" of the sphere S.] Normalization is necessary in order that we may consider the measure of a set as its statistical weight in further averages.

Now let $f(U)$ be a quantity which assumes a definite value in each of the stationary states U of our family. We can consider

$$f(U) = f(\alpha_1 U_1 + \cdots + \alpha_m U_m)$$

as a function of the variables $\alpha_1, \alpha_2, \cdots, \alpha_m$. According to the measure we have adopted, the mean value $<f(U)>$ of the quantity $f(U)$ is given by the integral

$$<f(U)> = \int_S dS \int_0^{2\pi} \cdots \int_0^{2\pi} f\left(\sum_{k=1}^{m} r_k e^{i\varphi_k} U_k\right) d\varphi_1 \cdots d\varphi_m,$$

which can, of course, be written more briefly in the form

$$<f(U)> = \int_{S^*} f\left(\sum_{k=1}^{m} \alpha_k U_k\right) dS^*,$$

where S^* is the complex sphere (2), and dS^* is the "surface" element of this sphere. ($dS^* = \mu dS\, d\varphi_1 \cdots d\varphi_m$, where μ is the aforementioned normalizing factor.) In the following, we shall call a quantity of the form $f(U)$ a *phase function* of our system, and the mean value $<f(U)>$ which we have defined, will be called its *microcanonical average*. We have established in this way a particular uniform principle for the construction of mean values of phase functions — the principle of "microcanonical" averaging.

We know, however, that the physical quantities which characterize the state of a system in quantum physics, in general, are not phase functions. Such a quantity $\mathfrak{A}$ can assume different values in a given state U, and from quantum theory we can obtain the distribution law of these values. As we know from Chapter II, to the quantity $\mathfrak{A}$ there corresponds a certain (linear self-adjoint) operator $\mathcal{A}$ which acts on the function U. The mathematical expectation $\mathsf{E}_U\mathfrak{A}$ of $\mathfrak{A}$ in the state U is equal to the scalar product $(\mathcal{A}U, U)$. This rule, if we apply it to all quantities $\mathfrak{A}$ and all states U, gives us, as we saw in Chapter II, the distribution law of any physical quantity as well as its mathematical expectation.

Therefore, in general, we cannot consider the physical quantity $\mathfrak{A}$ to be a phase function, and hence cannot speak of microcanonical averages of such quantities. But the mathematical expectation $(\mathcal{A}U, U)$ of $\mathfrak{A}$ in the state U is always a phase function, so that we can speak of the microcanonical average $<(\mathfrak{A}U, U)>$ of this mathematical expectation:

$$<(\mathfrak{A}U, U)> = <\mathsf{E}_U\mathfrak{A}> = \int_{S^*} (\mathfrak{A}U, U)\, dS^*.$$

From now on we shall call this microcanonical average of the mathematical expectation of the quantity $\mathfrak{A}$ the microcanonical average of the *quantity* $\mathfrak{A}$ *itself*. (Such an extension of the concept of microcanonical averaging can hardly lead to an inconsistency, because, in the case of quantities which do not have a uniquely defined value in every state, the concept of microcanonical averaging has not been defined.) In this manner we obtain

$$(3) \qquad <\mathfrak{A}> = <\mathsf{E}_U\mathfrak{A}> = \int_{S^*} (\mathfrak{A}U, U)\, dS^*,$$

where $\mathfrak{A}$ is the operator corresponding to $\mathfrak{A}$. One can say that this relation defines the microcanonical average for phase functions whose values are not ordinary numbers but random variables.

The construction of the mathematical expectation of a random variable always consists of a certain averaging process. Therefore, in quantum statistics we shall at every step have to consider two averaging processes which are generally unrelated: the formation of the mathematical expectation of a random variable and the microcanonical average. It is extremely important never to confuse these two processes. In the sequel, the pair of angular brackets $< >$ will always denote a microcanonical average. The term "mean value" will also always denote a microcanonical average and must not be confused with the mathematical expectation of a random variable. Wherever necessary, we shall denote the mathematical expectation of $\mathfrak{A}$ in the state U, by the symbol $\mathsf{E}_U\mathfrak{A}$.

Now we reduce expression (3) for the microcanonical average of $\mathfrak{A}$ to a form which is considerably more convenient for our later purposes. Since

$$U = \sum_{k=1}^{m} \alpha_k U_k,$$

then by the linearity of the operator $\mathfrak{A}$,

$$(\mathfrak{A}U, U) = (\sum_{k=1}^{m} \alpha_k \mathfrak{A}U_k, \sum_{l=1}^{m} \alpha_l U_l).$$

Whence, by the known properties of the scalar product (see II, §1)

$$(\mathfrak{A}U, U) = \sum_{l,k=1}^{m} \alpha_k \alpha_l^* (\mathfrak{A}U_k, U_l).$$

Therefore, (3) gives

$$<\mathfrak{A}> = \sum_{l,k=1}^{m} (\mathfrak{A}U_k, U_l) \int_{S^*} \alpha_k \alpha_l^*\, dS^*.$$

For $k \neq l$,

$$\int_{S^*} \alpha_k \alpha_l^* \, dS^* = \mu(2\pi)^{m-2} \int_S r_k r_l \, dS \int_0^{2\pi} \int_0^{2\pi} e^{i(\varphi_k - \varphi_l)} \, d\varphi_k \, d\varphi_l = 0;$$

and for $k = l$,

$$\int_{S^*} |\alpha_k|^2 \, dS^* = (1/m) \int_{S^*} \left\{ \sum_{k=1}^{m} |\alpha_k|^2 \right\} dS^* = (1/m) \int_{S^*} dS^* = 1/m;$$

since, according to our normalization, the measure of the complex sphere S^* is unity. Thus, we obtain

$$\text{(4)} \qquad <\mathfrak{A}> = (1/m) \sum_{k=1}^{m} (\mathfrak{a} U_k, U_k).$$

This means that the microcanonical average of any physical quantity is simply equal to the arithmetic average of the mathematical expectations of this quantity over all the states of some orthonormal system. Thus, microcanonical averaging is equivalent to averaging over an orthonormal basis in which identical weights are ascribed to each term.

We have based our expression for the microcanonical average on a definite linear orthonormal basis $U_1, U_2, \cdots, U_m$ of the manifold $\mathfrak{M}_E$. It is obvious, however, that to be a meaningful expression, it must be independent of the chosen basis. We now demonstrate this independence. (As a matter of fact, this independence follows from the invariance, with respect to the choice of orthonormal basis, of the measure introduced on p. 76.) If $V_1, V_2, \cdots, V_m$ is another linear basis of the manifold $\mathfrak{M}_E$, then (II, §3)

$$V_k = \sum_i \lambda_{ik} U_i \qquad (1 \le k \le m),$$

where the numbers λ_{ik} constitute a unitary matrix, and the summation here and in the following extends from 1 to m. Thus,

$$(\mathfrak{a} V_k, V_k) = (\sum_i \lambda_{ik} \mathfrak{a} U_i, \sum_j \lambda_{jk} U_j)$$
$$= \sum_{i,j} \lambda_{ik} \lambda_{jk}^* (\mathfrak{a} U_i, U_j);$$

and hence,

$$\sum_k (\mathfrak{a} V_k, V_k) = \sum_{i,j} (\sum_k \lambda_{ik} \lambda_{jk}^*)(\mathfrak{a} U_i, U_j).$$

But, by the unitary property of the matrix λ_{ik}, we have

$$\sum_k \lambda_{ik} \lambda_{jk}^* = \delta_{ij} \qquad (1 \le i, j \le m).$$

Thus,

$$\sum_k (\mathfrak{a} V_k, V_k) = \sum_{i,j} \delta_{ij} (\mathfrak{a} U_i, U_j) = \sum_i (\mathfrak{a} U_i, U_i).$$

This equality shows that (4) is independent of the chosen basis.

§3. Complete, symmetric and antisymmetric statistics

The principle of microcanonical averaging, which we defined in §2, is obviously based on the idea of equal participation in the formation of mean values of all the eigenfunctions of the operator $\mathcal{H}$ corresponding to a given eigenvalue E. This was also the main idea in classical statistical mechanics. There too the microcanonical average was extended in a uniform manner over all points of phase space belonging to a given "surface of constant energy". In classical mechanics this idea was based on the assumption that the equations of motion had no other single-valued integrals besides the energy integral (see Introduction, §2). Thus, if there exists an integral I of the equations of motion independent of the energy integral, and if in a given system $I = C$, then only a subset of the set of points constituting the "energy surface" can actually represent states of the system. This subset constitutes a manifold of a smaller number of dimensions and is characterized by the energy of the system and the equation $I = C$. In this case the averaging in which all points of the "energy surface" share equally, will, as a rule, lead to incorrect results because in this averaging states which are inaccessible to the system would play an overwhelming role. Moreover, the fact that the methods of classical statistical mechanics, where applicable (i.e., where the transition to quantum physics is not required) always lead to results which agree well with experiment, doubtless shows that in these problems such "inhibiting" integrals are actually absent.

In quantum physics a similar question arises. If we choose a method of averaging in which all stationary states of the system, corresponding to the same value of its energy, receive an equal weight, then we must assume that nothing prevents our system from actually finding itself in any of these states, i.e., that all of these states are *accessible* to the system.

However, for the most important and frequently encountered systems of quantum physics, this assumption is incorrect. Such systems, as a rule, cannot attain all the states which are described by the eigenfunctions U of the operator $\mathcal{H}$, corresponding to its eigenvalue E. Only a very insignificant fraction of these states can be attained. A valid averaging method for such a system must therefore be significantly different from the microcanonical one, as we have defined it above. In a great many physically important cases this new averaging method leads to results which are substantially different from the microcanonical averages. We must now consider in detail how this comes about.

Let our system consist of n identical and completely indistinguishable particles. We denote by (q_i , p_i) the set of all Hamiltonian variables of the ith particle. The Hamiltonian function of the system, due to the indistinguishability of the particles, must be a symmetric function of the variables (q_1 , p_1), (q_2 , p_2), $\cdots$, (q_n , p_n). (We also allow the possibility of interaction between the particles, and even allow the Hamiltonian function to depend explicitly on the time.) Clearly, the operator $\mathcal{H}$ of the system will exhibit the same symmetry with regard to any pair of particles. Let the function

$$U = U(q_1, q_2, \cdots, q_n, t)$$

satisfy the Schrödinger equation

$$i\hbar\, \partial U/\partial t = \mathcal{H}U. \tag{5}$$

We show that in this case the function

$$\hat{U} = U(q_2, q_1, \cdots, q_n, t),$$

obtained from the function U by the permutation of the first two particles, is also a solution of equation (5).

We denote by $\mathcal{P}$ the operator which transforms any function

$$f(q_1, q_2, \cdots, q_n, t)$$

into $f(q_2, q_1, \cdots, q_n, t)$ (the permutation operator of the first two particles), and we analyze in more detail the meaning of the postulated symmetry of the operator $\mathcal{H}$. To this end, we shall have to define the meaning of the permutation of the first two particles in some operator $\mathcal{A}$. Let this permutation transform $\mathcal{A}$ into an operator $\hat{\mathcal{A}}$. (We cannot denote this new operator by $\mathcal{P}\mathcal{A}$, as would seem natural, because this symbol has another meaning for us.) Suppose now that we wish to permute the first two particles in the expression $\mathcal{A}U$. It is natural to think that for this purpose we must permute them both in the operator $\mathcal{A}$ (the result will be $\hat{\mathcal{A}}$) and in the function U (the result will be $\mathcal{P}U$). Therefore

$$\mathcal{P}\mathcal{A}U = \hat{\mathcal{A}}\mathcal{P}U.$$

Setting $\mathcal{P}U = V$, whence $U = \mathcal{P}V$, we find

$$\hat{\mathcal{A}}V = \mathcal{P}\mathcal{A}\mathcal{P}V;$$

or, in terms of operators alone,

$$\hat{\mathcal{A}} = \mathcal{P}\mathcal{A}\mathcal{P}.$$

This gives us a natural definition of the operator $\hat{\mathcal{A}}$. The symmetry of the operator $\mathcal{A}$ is expressed by the requirement $\hat{\mathcal{A}} = \mathcal{A}$, or

$$\mathcal{P}\mathcal{A}\mathcal{P} = \mathcal{A},$$

whence

$$\mathcal{P}\mathcal{P}\mathcal{A}\mathcal{P} = \mathcal{A}\mathcal{P} = \mathcal{P}\mathcal{A},$$

i.e., a symmetric operator must commute with the operator $\mathcal{P}$ of the permutation of two particles (and hence also with the operator of a permutation of the particles among themselves).

In particular, the symmetry of the operator $\mathcal{H}$ implies

$$\mathcal{P}\mathcal{H} = \mathcal{H}\mathcal{P},$$

i.e., the operators $\mathcal{P}$ and $\mathcal{H}$ commute with one another. Thus, from (5) it follows that

$$i\hbar\, \partial\hat{U}/\partial t = \mathcal{P}(i\hbar\, \partial U/\partial t) = \mathcal{P}\mathcal{H}U = \mathcal{H}\mathcal{P}U = \mathcal{H}\hat{U},$$

which proves our assertion.

Therefore, the function $\hat{U} = \mathcal{P}U$ will be a solution of the Schrödinger equation, and so also will the function

$$\psi_t = \psi(q_1, \cdots, q_n, t) = U - \hat{U}.$$

Let us suppose now, that at $t = 0$ the function U is symmetric with respect to q_1 and q_2, i.e., that

$$U(q_1, q_2, \cdots, q_n, 0) = U(q_2, q_1, \cdots, q_n, 0)$$

identically in the $q_1, q_2, \cdots, q_n$. In other words, we have $\psi_0 = 0$ identically in the $q_1, q_2, \cdots, q_n$. It follows that the norm of the function ψ_0 is also zero. Since, for functions satisfying the Schrödinger equation, the norm cannot vary with time (see II, §4), the function ψ_t has a vanishing norm for all t, from which it follows that $\psi_t = 0$ for all $t, q_1, q_2, \cdots, q_n$. This means that

$$U(q_1, q_2, \cdots, q_n, t) = U(q_2, q_1, \cdots, q_n, t)$$

for all values of these same variables.

Thus, every function which satisfies the Schrödinger equation and is symmetric with respect to a certain pair of particles at a certain (arbitrary) value of t, preserves this symmetry for all other (preceding or following) values of t. If, therefore, the state of a system at $t = 0$ is described by an eigenfunction which is symmetric with respect to a pair of particles, then no state which lacks this symmetry is accessible to the system, either in the past or in the future.

From the above it follows that if an eigenfunction

$$U = U(q_1, q_2, \cdots, q_n, t)$$

describing a state of the system, is symmetric with respect to any pair of particles at $t = 0$ (i.e., if it is a symmetric function of the variables $q_1, q_2, \cdots, q_n$), then it preserves this property at all other (preceding or following) times. For such a system only those states are accessible which are described by eigenfunctions symmetric with respect to all the variables q_i. Obviously, if the mean values of physical quantities are to have real meaning for systems of this kind, the averaging must be carried out only over symmetric eigenfunctions, because all other eigenfunctions describe states which are fundamentally inaccessible to the system. Therefore, taking these other functions into account in the formation of mean values could lead to results which are incompatible with physical reality.

As we know, the eigenfunctions corresponding to a given eigenvalue E of the operator $\mathcal{H}$ generate a linear manifold $\mathfrak{M}_E$, whose dimension we shall denote by m. The symmetric functions that belong to the manifold $\mathfrak{M}_E$ also generate a certain linear manifold $\mathfrak{S}_E$, whose dimension s is, in general, less than m. If we know that our system can be found only in states which are described by symmetric eigenfunctions, then for such a system, obviously, the averaging must extend over the manifold $\mathfrak{S}_E$, and not over the manifold $\mathfrak{M}_E$. In all other respects, the considerations which led us to the choice of a natural principle of averaging over the manifold $\mathfrak{M}_E$ remain valid in the new case. We choose an orthonormal linear basis S_1, $S_2, \cdots, S_s$ of the manifold $\mathfrak{S}_E$ and, in analogy with formula (4) of §2, we define the microcanonical average of the quantity $\mathfrak{A}$ by means of the relation

$$<\mathfrak{A}> = (1/s)\sum_{k=1}^{s} (\mathcal{A}S_k, S_k), \tag{6}$$

where $\mathcal{A}$ is the operator assigned to $\mathfrak{A}$.

If all of the possible stationary states of a system are described by symmetric eigenfunctions, then we shall say that this system is subject to *symmetric statistics*. For such a system, the microcanonical averages of all physical quantities must be computed from (6). Along with systems of this kind, we shall also consider systems whose statistics bear an antisymmetric character. A function

$$U = U(q_1, q_2, \cdots, q_n)$$

is called antisymmetric with respect to the variables q_1 and q_2 if

$$\mathcal{P}U = U(q_2, q_1, \cdots, q_n) = -U$$

identically, i.e., if interchanging the variables q_1 and q_2 causes only a change in the sign of the function. If q_i is understood to be the set of generalized coordinates which describe the position of the ith particle, then it is easy to see that antisymmetry with respect to a given pair of particles is, like symmetry, invariant with respect to the Schrödinger equation. From this it follows that if the state of a system is described at any time t by an eigenfunction antisymmetric with respect to any pair of particles, then the same description will obtain at all preceding and following times. For such a system only states described by antisymmetric eigenfunctions are accessible. In complete analogy with the above, we must conclude that for systems of this type *antisymmetric statistics* hold: All averages must be carried out over the linear manifold $\mathfrak{A}_E$ of antisymmetric eigenfunctions (comprising, like $\mathfrak{S}_E$, only a part of the manifold $\mathfrak{M}_E$).

All of the above still does not answer the question of whether we must consider the statistics, to which the system is subject, as a special property of the component particles or as a mere accident *for the chosen system.* For example, let the system obey symmetric statistics. Then will every other system, composed of particles of the same nature, under any conditions necessarily obey symmetric statistics, too?

Both experiments and many theoretical considerations, which we cannot discuss here, tell us that this is precisely the case. The kind of statistics obeyed by a certain system does not depend on the conditions in which it is found, nor on any random effects, but is entirely determined by the nature of the particles which constitute the system. The elementary material particles (electrons, protons, neutrons) always obey antisymmetric statistics. Photons, on the other hand, are always governed by symmetric statistics. The type of statistics obeyed by material particles of a more complex structure is determined by the number of elementary particles which comprise a single particle of the complex type. The statistics will be symmetric or antisymmetric depending on the evenness or oddness of this number. Finally, both experimental and theoretical considerations lead us to say that for systems which are composed of free "non-localized" particles (i.e., those not bound to a definite place) no other type of statistics, besides symmetric and antisymmetric, is encountered. If, on the contrary, the particles are localized in space, then they lose their indistinguishability. (The location of two particles at different places makes them distinct and enables us to distinguish between them.) In this case even the statistics which we originally defined can be valid, i.e., those in which averages are extended over the entire manifold $\mathfrak{M}_E$. We shall therefore call them *complete statistics.*

Thus, in quantum statistical physics we must invariably consider three distinct statistical schemes — complete, symmetric and antisymmetric. For each case we must select the scheme which is characteristic of the type of particle being studied. However, the general development of the theory for all three schemes is carried out along parallel lines.

We ascribe to each physical system, which is composed of particles of a single type, an "index of symmetry" σ. This index σ equals 0, 1, or -1, depending on whether the particles of the system are subject to complete, symmetric or antisymmetric statistics, respectively. Because the type of statistics, as we know, remains constant in time, we can consider σ to be a peculiar "integral" of the Schrödinger equation. Just as in classical mechanics each new single-valued integral of the equations of motion reduces the manifold over which averages must be extended, so here too the presence of the "integral" σ compels us (if $\sigma \neq 0$) to substitute for the manifold $\mathfrak{M}_E$ the reduced manifold $\mathfrak{S}_E$ or $\mathfrak{A}_E$ for the purpose of averaging.

§4. Construction of the fundamental linear basis

Suppose we have a system, composed of N identical particles of arbitrary structure, occupying a finite volume V. Let

$$u_1(x),\ u_2(x),\ \cdots,\ u_n(x),\ \cdots \tag{7}$$

be a complete orthonormal system (linear basis) of eigenfunctions of the energy operator of an individual particle; where, for brevity, we denote by x the set of coordinates which define the position of the particle. We number them so that the eigenvalues (energy levels) will form a non-decreasing sequence: If the eigenfunction $u_r(x)$ corresponds to the eigenvalue ε_r, then

$$\varepsilon_1 \leq \varepsilon_2 \leq \cdots \leq \varepsilon_r \leq \cdots .$$

In the sequence $\varepsilon_1, \varepsilon_2, \cdots, \varepsilon_r, \cdots$ each energy level is repeated a number of times equal to its multiplicity (degree of degeneracy), i.e., a number of times equal to the number of functions in the sequence (7) which correspond to this level.

Now we consider the function

$$U = u_{r_1}(q_1)u_{r_2}(q_2) \cdots u_{r_N}(q_N), \tag{8}$$

where q_i denotes the set of coordinates of the position of the ith particle, and $r_1, r_2, \cdots, r_N$ are arbitrary integers. Hence, U is a function of the positional coordinates of all N of the particles which compose the system (or, equivalently, of the positional coordinates of the system itself).

THEOREM 1. *The function U, defined by* (8), *is an eigenfunction of the energy operator $\mathcal{H}$ of the system, corresponding to the eigenvalue*

$$E = \varepsilon_{r_1} + \varepsilon_{r_2} + \cdots + \varepsilon_{r_N} \tag{9}$$

of this operator.

Proof. Denoting by $\mathcal{H}_i$ the energy operator of the ith particle, we have

$$\mathcal{H} = \textstyle\sum_{i=1}^{N} \mathcal{H}_i ,$$

and hence

$$\mathcal{H}U = \textstyle\sum_{i=1}^{N} \mathcal{H}_i U.$$

But, since the operator $\mathcal{H}_i$ operates only on the function $u_{r_i}(q_i)$

$$\mathcal{H}_i U = u_{r_1}(q_1) \cdots u_{r_{i-1}}(q_{i-1})[\mathcal{H}_i u_{r_i}(q_i)]u_{r_{i+1}}(q_{i+1}) \cdots u_{r_N}(q_N);$$

and since $u_{r_i}(q_i)$ is an eigenfunction of the operator $\mathcal{H}_i$, belonging to the eigenvalue ε_{r_i}, we have

$$\mathcal{H}_i u_{r_i}(q_i) = \varepsilon_{r_i} u_{r_i}(q_i).$$

Hence,

$$\mathcal{H}_i U = \varepsilon_{r_i} U,$$

and

$$\mathcal{H} U = \textstyle\sum_{i=1}^N \mathcal{H}_i U = (\sum_{i=1}^N \varepsilon_{r_i}) U = EU,$$

q.e.d.

If the energy level E of the system is assigned beforehand, then any choice of integers r_i ($i = 1, 2, \cdots, N$), which satisfy relation (9), gives us an eigenfunction U of the operator $\mathcal{H}$, corresponding to this level.

THEOREM 2. *The set of all functions of the form* (8), *obtained for all values of the integers* $r_1, r_2, \cdots, r_N$, *which satisfy relation* (9), *is a linear orthonormal basis of the manifold* $\mathfrak{M}_E$ *of eigenfunctions of the energy operator* $\mathcal{H}$ *of the given system, corresponding to its eigenvalue* E.

Proof. We show first that the set of all functions of the form (8), obtained for all possible values of the integers $r_1, r_2, \cdots, r_N$, *satisfying relation* (9) *or not*, represents a complete orthonormal system of functions. The orthogonality and normality of this system follow in an obvious way from the corresponding properties of system (7), and the proof requires showing only the *completeness* of the system (8). This proof is most easily carried out with the help of induction on the number of particles N. As a matter of fact, for $N = 1$, the completeness of system (8) coincides with the assumed completeness of system (7). Therefore, take $N > 1$, and let our assertion be true for systems composed of $N - 1$ particles. Let

$$\varphi(q_1, q_2, \cdots, q_N)$$

be orthogonal to all functions of the form (8), so that for any values of the integers $r_1, r_2, \cdots, r_N$,

$$I = \int \varphi^*(q_1, \cdots, q_N) u_{r_1}(q_1) \cdots u_{r_N}(q_N)\, dq_1 \cdots dq_N = 0.$$

Setting

$$\int \varphi^*(q_1, \cdots, q_N) u_{r_1}(q_1) \cdots u_{r_{N-1}}(q_{N-1})\, dq_1 \cdots dq_{N-1} = \psi_{r_1 \cdots r_{N-1}}(q_N), \tag{10}$$

we have

$$I = \int_{-\infty}^{\infty} \psi_{r_1 \cdots r_{N-1}}(q_N) u_{r_N}(q_N) \, dq_N = 0.$$

Because this equality holds for any r_N, by the completeness of system (7)

$$\psi_{r_1 \cdots r_{N-1}}(q_N) \equiv 0$$

for any $r_1, r_2, \cdots, r_{N-1}$. By (10) this gives for any $r_1, r_2, \cdots, r_{N-1}$ and any fixed q_N,

$$\int \varphi^*(q_1, \cdots, q_N) u_{r_1}(q_1) \cdots u_{r_{N-1}}(q_{N-1}) \, dq_1 \cdots dq_{N-1} = 0.$$

It follows by mathematical induction that

$$\varphi(q_1, \cdots, q_N) = 0$$

for any q_N, and identically in the $q_1, \cdots, q_{N-1}$. In other words, this equation is satisfied identically in all N variables, which proves our assertion.

It is now easy to show that functions of the form (8), in which the numbers $r_1, \cdots, r_N$ are related by (9), constitute a linear basis of the manifold $\mathfrak{M}_E$. In fact, let $V = V(q_1, \cdots, q_N)$ be any eigenfunction of the operator $\mathcal{H}$, corresponding to the eigenvalue E. By the established completeness of the system of functions of the form (8) we can write

$$V = \sum_i c_i U_i,$$

where the summation extends over the whole set of functions of the form (8). Use of the general rule for determining Fourier coefficients (II, §3) gives

$$c_i = (V, U_i) \qquad (i = 1, 2, \cdots).$$

The function U_i, like V, is by Theorem 1, an eigenfunction of the operator $\mathcal{H}$. If the eigenvalue $\varepsilon_{r_1} + \varepsilon_{r_2} + \cdots + \varepsilon_{r_N}$, corresponding to U_i, is not equal to E, i.e., if it does not satisfy (9), then V and U_i belong to different eigenvalues of this operator and are thus mutually orthogonal (II, §3), so that $c_i = 0$. Hence, the function V is represented as a linear combination of functions of the form (8), in which the numbers r_i are related by (9). This is precisely what we were required to show.

Therefore, the set of functions of the form (8), for which $\varepsilon_{r_1} + \varepsilon_{r_2} + \cdots + \varepsilon_{r_N} = E$, is a linear orthonormal basis of the manifold $\mathfrak{M}_E$. We shall henceforth always call these functions the *fundamental eigenfunctions*, and the states described by them the *fundamental states* of the system. The set of fundamental states of a system is uniquely defined by the choice of the original orthogonal set (7) and by the assignment of the number of particles N and the energy E of the system. In particular, if the structure of the constituent particles is known, then the *number* m of fundamental states of the system is a function of N and E, which we shall denote by $\Omega(N, E)$. We call it the *structure function* of the system. This function, as we shall see, plays a leading role in our theory. Denoting the fundamental eigenfunctions corresponding to the energy level E by $U_1, U_2, \cdots, U_m$, we have for any physical quantity $\mathfrak{A}$ *in the case of complete statistics*,

$$<\mathfrak{A}> = [\Omega(N, E)]^{-1} \sum_{k=1}^{m} (\mathcal{A}U_k, U_k).$$

(This follows from §2.)

We shall presently prove that the above linear basis, composed of fundamental eigenfunctions, is especially convenient for the subsequent development of the theory. However, we return now to the consideration of symmetric statistics.

In the expression (8) for a fundamental function U the indices of the variables $q_1, q_2, \cdots, q_N$ comprise the sequence of integers from 1 to N. We generate an arbitrary permutation $\mathcal{P}$ among them, i.e., we transform the function U to the function

$$\mathcal{P}U = u_{r_1}(q_{i_1})u_{r_2}(q_{i_2}) \cdots u_{r_N}(q_{i_N}),$$

where $i_1, i_2, \cdots, i_N$ are the numbers of the sequence $1, 2, \cdots, N$, in an order defined by the permutation $\mathcal{P}$. The number of possible permutations is obviously equal to $N!$. We now define S such that

$$S = S(q_1, q_2, \cdots, q_N) = \sum_{\mathcal{P}} \mathcal{P}U,$$

where the summation extends over all permutations of the sequence 1, 2, $\cdots$, N that give different functions $\mathcal{P}U$. It is obvious that S is a symmetric function of the variables $q_1, q_2, \cdots, q_N$ and at the same time an eigenfunction of the operator $\mathcal{H}$ corresponding to the eigenvalue E.

This construction, which we have just carried out for one fundamental function U, can be carried out for each of the fundamental functions of our fundamental linear basis. From each construction we obtain a certain symmetric function S. Clearly, some of these functions can be identical. We denote by $S_1, S_2, \cdots, S_s$ the set of distinct eigenfunctions obtained in this way, and by $\mathfrak{S}_E$ the set of all symmetric functions of the manifold $\mathfrak{M}_E$, i.e., all symmetric eigenfunctions of the operator $\mathcal{H}$ corresponding to the eigenvalue E.

THEOREM 3. *The functions $S_1, S_2, \cdots, S_s$ constitute an orthogonal linear basis of the manifold $\mathfrak{S}_E$.*

Proof. 1. Suppose $m \neq n$, and

$$S_m = \sum_{\mathcal{P}} \mathcal{P}U_m, \qquad S_n = \sum_{\mathcal{P}'} \mathcal{P}'U_n. \tag{S}$$

Then

$$(S_m, S_n) = \sum_{\mathcal{P},\mathcal{P}'} (\mathcal{P}U_m, \mathcal{P}'U_n),$$

where the summation extends over all pairs $\mathcal{P}$, $\mathcal{P}'$ corresponding to permutations of the sequence 1, 2, $\cdots$, N which occur in (S). If any of the scalar products $(\mathcal{P}U_m, \mathcal{P}'U_n)$ are different from zero, then we must have $\mathcal{P}U_m = \mathcal{P}'U_n$, because $\mathcal{P}U_m$ and $\mathcal{P}'U_n$ are fundamental functions, i.e., members of the *orthogonal* system of functions (8). From $\mathcal{P}U_m = \mathcal{P}'U_n$ it follows that

$$U_m = \mathcal{P}^{-1}\mathcal{P}'U_n = \mathcal{Q}U_n,$$

where $\mathcal{Q} = \mathcal{P}^{-1}\mathcal{P}'$ is a permutation of the sequence 1, 2, $\cdots$, N. Thus, the function U_m is obtained from a certain permutation $\mathcal{Q}$ on the function U_n. This leads to the contradiction $S_m = S_n$, from which it follows that $(S_m, S_n) = 0$ for $m \neq n$, i.e., the functions $S_1, S_2, \cdots, S_s$ are pairwise orthogonal.

2. Let S be any function of the manifold $\mathfrak{S}_E$ (and hence also a function of the manifold $\mathfrak{M}_E$). By Theorem 2,

$$S = \sum_{k=1}^{m} \gamma_k U_k, \tag{11}$$

where the γ_k are complex numbers, and the U_k comprise the fundamental linear basis of the manifold $\mathfrak{M}_E$. Let some pair of functions U_k and U_l be related by $U_l = \mathcal{P}U_k$, where $\mathcal{P}$ is a permutation of the sequence 1, 2, $\cdots$, N. Then

$$\gamma_l = (S, U_l) = (S, \mathcal{P}U_k).$$

By the symmetry of the function S, we must have $\mathcal{P}S = S$, and hence

$$\gamma_l = (\mathcal{P}S, \mathcal{P}U_k).$$

If one represents scalar products in the form of integrals, then $(\mathcal{P}S, \mathcal{P}U_k)$ obviously differs from (S, U_k) only in the notation of variables in the integral, so that

$$\gamma_l = (S, U_k) = \gamma_k.$$

Thus, in the sum (11), all functions U_l of the form $\mathcal{P}U_k$ have the same coefficient, namely γ_k. The sum of the corresponding group of terms on the right side of (11) is therefore

$$\gamma_k \sum_{\mathcal{P}} \mathcal{P}U_k = \gamma_k S_j,$$

where S_j is one of the functions $S_1, S_2, \cdots, S_s$. Since the right side of (11) is composed entirely of such groups,

$$S = \sum_{j=1}^{s} \lambda_j S_j,$$

where the λ_j are complex numbers. Hence, the functions S_j constitute a linear basis of the manifold $\mathfrak{S}_E$. This proves Theorem 3.

When the particles of a system obey symmetric statistics we shall call the functions $S_1, S_2, \cdots, S_s$ (after normalization through multiplication by the appropriate normalizing factor) the fundamental eigenfunctions, and the states described by them the fundamental states of the system for the given energy level E. The number s of these fundamental states we shall denote, as before, by $\Omega(N, E)$ and call it the structure function of the system. The microcanonical average of a physical quantity $\mathfrak{A}$, when the averaging is carried out only over the symmetric (normalized) eigenfunctions, is expressed by the formula

$$<\mathfrak{A}> = [\Omega(N, E)]^{-1} \sum_{j=1}^{s} (\mathcal{A}S_j, S_j).$$

We now turn to the case of antisymmetric statistics. For every function U of the form (8) we form the function

$$A = A(q_1, q_2, \cdots, q_N) = \sum_{\mathcal{P}} \pm \mathcal{P}U,$$

where the summation extends over *all* permutations of the sequence 1, 2, $\cdots, N$, and the sign $+$ or $-$ is used depending on the evenness or oddness of the permutation $\mathcal{P}$. It is easy to see that the function A can be rewritten in the form of a determinant

$$A = \begin{vmatrix} u_{r_1}(q_1) & u_{r_1}(q_2) & \cdots & u_{r_1}(q_N) \\ u_{r_2}(q_1) & u_{r_2}(q_2) & \cdots & u_{r_2}(q_N) \\ \cdots & \cdots & \cdots & \cdots \\ u_{r_N}(q_1) & u_{r_N}(q_2) & \cdots & u_{r_N}(q_N) \end{vmatrix}. \tag{12}$$

However, we shall hardly ever need this expression. We carry out the construction indicated above for each of the functions U of the form (8) and denote (after normalization) by $A_1, A_2, \cdots, A_a$ the set of (antisymmetric) functions obtained in this way. The members of this set differ from one another and do not vanish. Let $\mathfrak{A}_E$ be the manifold of all antisymmetric functions of the manifold $\mathfrak{M}_E$. Then the following theorem is valid.

THEOREM 4. *The functions $A_1, A_2, \cdots, A_a$ constitute an orthogonal linear basis of the manifold $\mathfrak{A}_E$.*

It is not necessary to prove this theorem because the proof is analogous to that of Theorem 3. When the system consists of particles that obey antisymmetric statistics, we shall call the (normalized) functions $A_1, A_2, \cdots, A_a$ the fundamental eigenfunctions, and the states described by them the fundamental states of the system for the given energy level E. The number a of these fundamental states we again denote by $\Omega(N, E)$ and call it the structure function of the system. For the microcanonical average of a quantity $\mathfrak{A}$ in this case, we average over only the antisymmetric eigenfunctions. Hence,

$$<\mathfrak{A}> = [\Omega(N, E)]^{-1} \sum_{j=1}^{a} (\mathcal{A}A_j, A_j).$$

The fundamental states of a system which are described by the antisymmetric eigenfunctions A_j, possess one extremely important characteristic: If the system is in such a state, then the states of any two of its particles must differ from one another. In fact, if in the eigenfunction of such a state two of the indices r_i are equal, then the determinant (12) will have two identical rows, and will therefore vanish. Thus,

$$A(q_1, q_2, \cdots, q_N) = 0.$$

This type of function cannot be normalized and does not describe any real state of the system. The above rule is usually called the *Pauli principle.*

In the case of complete statistics, if the system is in a fundamental state, then the state of each of the particles composing the system is known. [The state of the ith particle is determined by the index r_i in the expression (8) for the fundamental state U.] In the case of symmetric or antisymmetric statistics such a statement would not be meaningful. For example, let

$$S = \sum_{\mathcal{P}} \mathcal{P}U$$

be one of the fundamental symmetric eigenfunctions. In the different states $\mathcal{P}U$ of the state S, the first particle, for instance, is described by the different functions $u_{r_j}(q_1)$, so that in the state S the first particle is not in any definite state. In the case of symmetric or antisymmetric statistics the only sensible question is *how many particles* are to be found in some particular state if we know the state of the overall system. We discuss this question in detail in the following section.

§5. Occupation numbers. Basic expressions of the structure functions

Suppose we are given a stationary state U of a system which obeys one of the three statistics. U, as usual, is an eigenfunction of the energy operator $\mathcal{H}$, corresponding to a definite eigenvalue (energy level) E of this operator. We can ask how many particles will be found in a state which is described by an eigenfunction $u_r(x)$ (corresponding to the energy level ε_r of a particle). The number $a_r = a_r(U)$ of such particles is, generally, a physical quantity whose value is not uniquely determined by the state U. The only characteristic determined by the state U is the distribution law of the number a_r and, in particular, its mathematical expectation $\mathsf{E}_U a_r$. But *in every fundamental state of the system the value of the number a_r is uniquely determined.* So, if we are concerned with complete statistics, then each of the fundamental eigenfunctions, according to formula (8) of §4, has the form

$$U = u_{r_1}(q_1)u_{r_2}(q_2) \cdots u_{r_N}(q_N). \tag{8}$$

This shows directly which of its possible states each particle occupies. The number a_r of particles in the state $u_r(x)$ is simply the number of indices r_i which equal r in expression (8). Clearly, this number has a definite value for each fundamental state. We are easily led to the same result in the case of the other two statistics. In these latter cases the fundamental functions are linear combinations of functions of the form (8), where in all the terms the number of indices r_i, equal to any given number, is the same, and hence, the number a_r is uniquely determined.

If the state U is not fundamental, then it can be represented in the form

$$U = \sum_{i=1}^{k} \alpha_i U_i ,$$

where the α_i are complex numbers. (For the purposes of normalization we shall suppose that $|\alpha_1|^2 + |\alpha_2|^2 + \cdots + |\alpha_k|^2 = 1$.) The U_i are the fundamental states and among them there will be fundamental states with different values of the number a_r, so that it is impossible to speak of any definite value of this number in the state U. One can only assert that if $a_r^{(i)}$ is the value of the number a_r in the fundamental state U_i, then $\sum_j |\alpha_j|^2$ is the probability that a_r will equal $a_r^{(i)}$ in the state U, where the summation extends over all j for which $a_r^{(j)} = a_r^{(i)}$. Thus, in particular,

$$\mathsf{E}_U a_r = \sum_{i=1}^{k} |\alpha_i|^2 a_r^{(i)}.$$

(We do not prove this statement as it will not be of further interest to us.)

In statistical physics the numbers a_r ($r = 1, 2, \cdots$), which represent the numbers of particles in the states described by eigenfunctions $u_r(x)$ ($r = 1, 2, \cdots$), are usually called "occupation numbers". It is obvious that if a system is composed of N particles and has an energy E, then, independently of the kind of statistics obeyed by the system, we must have

$$\sum_{r=1}^{\infty} a_r = N, \qquad \sum_{r=1}^{\infty} a_r \varepsilon_r = E. \tag{K}$$

The occupation numbers play a very important part in any description of statistical physics. The fact that for fundamental states these numbers take on uniquely determined values, makes the systems of fundamental eigenfunctions especially convenient linear bases of the manifolds over which averages must extend. (The particular manifold used, of course, depends on the type of statistics obeyed by the system.)

We have seen that in each of the three statistics, a definite set of occupation numbers $a_1, a_2, \cdots, a_r, \cdots$, which satisfy relation (K), corresponds to each fundamental state of the system. Now we pose the converse problem: Suppose we are given a set of non-negative integers a_r ($r = 1, 2, \cdots$), satisfying relation (K); then do fundamental states of the system exist which have the numbers a_r for occupation numbers, and, if so, how many of these fundamental states are there?

This problem, which is very important for the development of the statistical theory, is solved in different ways for the three statistical schemes.

1. *Complete Statistics.* Let $a_1, a_2, \cdots, a_r, \cdots$ be non-negative integers, related by (K). In order that a fundamental state (8) have the numbers a_r for its occupation numbers, it is necessary and sufficient that the function $u_r(x)$ occur a total of a_r times among the functions which are the factors of U. Since $\sum_{r=1}^{\infty} a_r = N$, it is obviously always possible to realize this requirement. The number of ways in which this can be accomplished is equal to the number of permutations of the set of N elements composed of subsets of $a_1, a_2, \cdots$ identical elements, i.e., it is equal to

$$N!/a_1!\,a_2! \cdots a_r!.$$

Thus, *in the case of complete statistics, to each solution* (in terms of non-negative integers a_r) *of the system of equations* (K), *there correspond*

$$N!/\prod_{r=1}^{\infty} a_r!$$

fundamental states having the numbers a_r for their occupation numbers.

2. *Symmetric Statistics.* In the expression

$$S = \sum_{\mathcal{P}} \mathcal{P}U$$

of the fundamental states for the case of symmetric statistics the terms of the right side differ from one another only by the order of the indices i_k in the product

$$u_{r_1}(q_{i_1})u_{r_2}(q_{i_2}) \cdots u_{r_N}(q_{i_N}). \tag{13}$$

The indices r_j are the same in all the terms, and hence the occupation numbers are also the same for all the terms. Clearly the converse also holds: All products of the form (13), which have a definite set of occupation numbers, can differ from one another only in the order of the indices i_k, and are obtained from one another by the corresponding permutation of these indices. This means that all such products constitute a single symmetric fundamental function S, which is thus uniquely determined by the given set of occupation numbers: *In the case of symmetric statistics to each solution* (in terms of non-negative integers a_r) *of the system of equations* (K), *there corresponds one fundamental state, having the numbers a_r for its occupation numbers.*

3. *Antisymmetric Statistics.* In this case for the fundamental functions,

we have the expression

$$A = \sum_{\mathcal{P}} \pm \mathcal{P}U,$$

where the + sign is to be used for even, and the − sign for odd permutations $\mathcal{P}$. Here too, the terms of the right side comprise the whole set of functions of the form (13) which have the same set of occupation numbers. Hence each solution of the system (K) cannot correspond to more than one antisymmetric state. However, we do not get a normalizable eigenfunction A for every set of occupation numbers — in some cases it turns out to vanish identically. Consider an arbitrary set of occupation numbers a_r $(r = 1, 2, \cdots)$. If each a_r is either 0 or 1, then in the product (13) all of the indices r_k are different, and so two such products, obtained from one another by some permutation of the indices i_k, cannot be the same. Thus in the sum $\sum_{\mathcal{P}} \pm \mathcal{P}U$, it is impossible for identical terms to enter, and A will be a normalizable eigenfunction. If, among the numbers a_r, there should be one greater than unity, then among the indices r_k in the product (13) some will be equal. Because the order of enumeration is immaterial, let us suppose that $r_1 = r_2$. Then the product (13) actually coincides with the product

$$u_{r_1}(q_{i_2})u_{r_2}(q_{i_1}) \cdots u_{r_N}(q_{i_N}),$$

which is obtained from it by permutation of the indices i_1 and i_2. Because each of these two products is obtained from the other by a simple transposition, they will have opposite signs in the sum which constitutes A, and will cancel one another. Since such a cancelling partner can be found for every member of this sum, $A = 0$.

Thus, *in the case of antisymmetric statistics, to each solution* (in terms of non-negative integers a_r) *of the system of equations* (K), *there corresponds:* 1) *one fundamental state, having the numbers* a_r *for its occupation numbers, if all* $a_r \leq 1$, or 2) *no such states if there is an* $a_r > 1$. It will be convenient for us to have an expression for the number of fundamental states corresponding to a given set of occupation numbers which will apply to all three statistical schemes. For this purpose we put, for $n = 1, 2, \cdots$,

$$C(n) = \begin{cases} n! & \text{for complete statistics,} \\ 1 & \text{for symmetric or antisymmetric statistics;} \end{cases}$$

and

$$\gamma(n) = \begin{cases} 1/n! & \text{for complete statistics,} \\ 1 & \text{for symmetric statistics,} \\ 1 \quad (n \leq 1) & \text{for antisymmetric statistics.} \\ 0 \quad (n > 1) & \end{cases}$$

Then, the number of fundamental states corresponding to a given set of numbers $a_r \geq 0$, satisfying the system of equations (K), is

$$C(N)\prod_{r=1}^{\infty} \gamma(a_r),$$

for any of the three statistics. We obtain from this, for the total number of fundamental states of the system corresponding to a given energy level E, the expression

$$\Omega(N, E) = C(N) \sum_{(K)} \prod_{r=1}^{\infty} \gamma(a_r), \tag{14}$$

where the summation on the right side extends over all systems of integers $a_1, a_2, \cdots, a_r, \cdots$ $(a_r \geq 0)$, satisfying equation (K).

As we have already remarked, the structure function $\Omega(N, E)$ plays a very important part in our subsequent development. Hence we shall have to study it with considerable care. Formula (14), which relates the function $\Omega(N, E)$ to the solutions of the system of equations (K), will serve as a starting point for our investigation.

For any statistics the occupation numbers a_r are uniquely determined for each fundamental state and can be considered as functions of this state. Hence, for any energy level E they have definite (microcanonical) mean values $<a_r>$. We shall see further on that these mean values of the occupation numbers are of fundamental importance for all the computational formulas of statistical physics. The determination of their asymptotic expressions therefore constitutes one of the basic mathematical problems of the theory and the following two chapters will be devoted mainly to this problem. However, we pause here to discuss certain general considerations which, to a large extent, clarify the importance of the numbers $<a_r>$ in the asymptotic computations of statistical physics.

A large fraction of the most important physical quantities $\mathfrak{A}$, which characterize the state of a system composed of N identical particles, have a certain special form: Such a quantity $\mathfrak{A}$ is a sum of N quantities $\mathfrak{A}_1, \mathfrak{A}_2, \cdots, \mathfrak{A}_N$, each of which depends (in the very same way) only on the state of the particle with the corresponding number. Therefore, in the simplest case, when the state of the system is described by the function

$$U = u_{r_1}(q_1)u_{r_2}(q_2) \cdots u_{r_N}(q_N),$$

the quantity $\mathfrak{A}_k$ depends only on the number r_k, i.e., on the index of the eigenfunction $u_{r_k}(q_k)$ which describes the state of the kth particle. To the decomposition

$$\mathfrak{A} = \mathfrak{A}_1 + \mathfrak{A}_2 + \cdots + \mathfrak{A}_N$$

of the quantity $\mathfrak{A}$ there corresponds, of course, the decomposition

$$\mathcal{A} = \mathcal{A}_1 + \mathcal{A}_2 + \cdots + \mathcal{A}_N$$

of the corresponding operator. In this latter decomposition the operator

$\mathfrak{A}_k$ $(k = 1, 2, \cdots, N)$ acts only on the eigenfunction of the kth particle, so that

$$\mathfrak{A}_k U = u_{r_1}(q_1) \cdots u_{r_{k-1}}(q_{k-1})[\mathfrak{A}_k u_{r_k}(q_k)]u_{r_{k+1}}(q_{k+1}) \cdots u_{r_N}(q_N).$$

Quantities $\mathfrak{A}$ of the above type will be called *sum* functions. We now consider how to construct the mean values of these sum functions.

If, as we have assumed, the dependence of the quantity $\mathfrak{A}_k$ on the state of the kth particle is the same for all k, then, in particular, the mathematical expectation of the quantity $\mathfrak{A}_k$ in some state $u_r(q_k)$ of the kth particle does not depend on k, but only on r. We denote this mathematical expectation by λ_r. Thus the set of numbers λ_r depends only on the structure of the particles and on the choice of the elementary quantity $\mathfrak{A}_k$. These numbers are not related to the system and retain their meaning and their values even when we consider an individual particle instead of a system.

Now let our system be in a fundamental state U with occupation numbers $a_1, a_2, \cdots, a_r, \cdots$. Then, first of all,

$$\mathsf{E}_U \mathfrak{A} = \textstyle\sum_{k=1}^{N} \mathsf{E}_U \mathfrak{A}_k . \tag{15}$$

It is obvious that $\mathsf{E}_U \mathfrak{A}_k = \lambda_r$, if the kth particle is in the state $u_r(q_k)$. Therefore, in the sum (15) the term λ_r appears as many times as there are particles in state $u_r(q_k)$, i.e., a_r times. Whence,

$$\mathsf{E}_U \mathfrak{A} = \textstyle\sum_{r=1}^{\infty} a_r \lambda_r ,$$

and consequently,

$$<\mathfrak{A}> = <\mathsf{E}_U \mathfrak{A}> = \textstyle\sum_{r=1}^{\infty} <a_r> \lambda_r . \tag{16}$$

This simple relation shows that a knowledge of the mean values of the occupation numbers allows us to write down immediately the mean value of any sum function $\mathfrak{A}$. (It is necessary to recall that the numbers λ_r do not depend on the state of the system and they can be calculated once and for all for particles of a given structure and for a given choice of the quantity $\mathfrak{A}_k$.) This use of the occupation numbers is the chief reason that the first problem considered in every statistical theory is that of finding convenient asymptotic expressions for their mean values.

§6. On the suitability of microcanonical averages

In §1 we spoke in some detail about the fact that physical quantities, which characterize the state of a system from the phenomenological point of view, must, in the statistical theory, depend symmetrically on all of the particles composing the system. The mean value of such a quantity, to really exist, i.e., to really have a physical meaning in a given experiment, must have a value close to that one of the values given by the statistical theory which is appropriate for the particular experiment. We pointed out in detail in §1 that for this purpose the mean values given by our theory must be *suitable*; that is, in the overwhelming majority of those states over which averages are extended a quantity must assume values extremely close to its mean value.

For a given total energy E, we have defined the mean value of the quantity $\mathfrak{A}$ by

$$(17) \qquad <\mathfrak{A}> \;=\; <\mathsf{E}_U\mathfrak{A}> \;=\; [\Omega(N, E)]^{-1} \textstyle\sum_{k=1}^{m} \mathsf{E}_{U_k}\mathfrak{A},$$

where $m = \Omega(N, E)$ is the number of fundamental states U_k of the system for the energy level E, and where the summation extends over all such states. The system, of course, can obey any of the three types of statistics we have considered.

The simplest and most important type of quantity $\mathfrak{A}$ which depends symmetrically on all the particles composing the system is obviously the sum function considered at the end of the preceding section. This function is a sum of N quantities, each depending on the state of a single particle and all having the same functional dependence. In the present section we shall always write

$$\mathfrak{A} = \textstyle\sum_{i=1}^{N} \mathfrak{A}_i$$

for such a sum function. Because the dependence of the quantity $\mathfrak{A}_i$ on the state of the ith particle is the same for all i, the mean value $<\mathfrak{A}_i> \; = \alpha$ does not depend on i. Obviously we have

$$<\mathfrak{A}> \; = N\alpha,$$

so that, as long as $\alpha \neq 0$, we can always think of $<\mathfrak{A}>$ as a large number (of the order of N). We can therefore think of the value of the quantity $\mathfrak{A}$ as near $<\mathfrak{A}>$, if

$$(18) \qquad | \mathfrak{A} - N\alpha | < \epsilon N,$$

where ϵ is a small positive number; because if (18) holds, the relative error in the equation $\mathfrak{A} \approx \; <\mathfrak{A}>$ is smaller than $| \epsilon/\alpha |$.

We recall that if the given system is in a fundamental state, we cannot say whether inequality (18) is satisfied or not, because in general the quantity $\mathfrak{A}$ is not uniquely determined in the state U. If the system is in state U, all that is determined is the probability

$$\mathsf{P}_U\{ | \mathfrak{A} - N\alpha | < \epsilon N\}$$

that inequality (18) is satisfied.

Now let M_δ denote the measure (in accordance with the definition of measure given in §2) of the set of all eigenfunctions U of the manifold $\mathfrak{M}_E$ for which

$$\mathsf{P}_U\{ | \mathfrak{A} - N\alpha | < \epsilon N\} < 1 - \delta;$$

or, equivalently,

$$\mathsf{P}_U\{\,|\,\mathfrak{A} - N\alpha\,|\, > \epsilon N\} > \delta, \tag{19}$$

where δ is any (small) positive number. Then assuming $U = \sum_{k=1}^{m} \alpha_k U_k$, we have

$$\int_{S^*} \mathsf{P}_U\{|\mathfrak{A} - N\alpha| > \epsilon N\}\, dS^* \geq \delta M_\delta,$$

where S^* is the complex sphere (2) of §2. Hence,

$$M_\delta \leq (1/\delta) \int_{S^*} \mathsf{P}_U\{|\mathfrak{A} - N\alpha| > \epsilon N\}\, dS^*.$$

But, by the Chebyshev inequality,

$$\mathsf{P}_U\{\,|\,\mathfrak{A} - N\alpha\,|\, > \epsilon N\} \leq \mathsf{E}_U\{(\mathfrak{A} - \alpha N)^2\}/\epsilon^2 N^2;$$

whence,

$$\begin{aligned} M_\delta &\leq (1/\delta\epsilon^2 N^2) \int_{S^*} \mathsf{E}_U\{(\mathfrak{A} - \alpha N)^2\}\, dS^* \\ &= <(\mathfrak{A} - \alpha N)^2>/\delta\epsilon^2 N^2. \end{aligned} \tag{20}$$

M_δ denotes the measure of the set of those stationary states U of the manifold $\mathfrak{M}_E$ for which inequality (19) holds. For all other states we can expect, with a probability exceeding $1 - \delta$, that inequality (18) will hold, i.e., that the approximate equality $\mathfrak{A} \approx <\mathfrak{A}>$ will hold. Thus, if M_δ becomes small for small ϵ and δ, then in a large majority of the states U we can expect, with overwhelming probability, that the quantity $\mathfrak{A}$ has a value near $<\mathfrak{A}>$. This is just what we call the suitability of the mean value $<\mathfrak{A}>$. As inequality (20) shows, this suitability will be guaranteed if, for small ϵ and δ, the quantity

$$<(\mathfrak{A} - \alpha N)^2>/\delta\epsilon^2 N^2$$

remains sufficiently small.

For brevity we denote the mean value $<(\mathfrak{A} - \alpha N)^2>$ by $D(\mathfrak{A})$ and call it the *microcanonical dispersion* of the quantity $\mathfrak{A}$. In the above calculation we have nowhere assumed that $\mathfrak{A}$ was a sum function. Now we determine the special form of the microcanonical dispersion which is appropriate for sum functions. Obviously, we have

$$\begin{aligned} D(\mathfrak{A}) &= <(\mathfrak{A} - \alpha N)^2> \\ &= <\mathfrak{A}^2> - \alpha^2 N^2 \\ &= <(\textstyle\sum_{i=1}^{N} \mathfrak{A}_i)^2> - \alpha^2 N^2. \end{aligned} \tag{21}$$

But for any state U of the system

$$\mathsf{E}_U(\mathfrak{A}^2) = \sum_{i=1}^{N} \mathsf{E}_U(\mathfrak{A}_i^2) + 2\sum_{\substack{i,j=1 \\ i<j}}^{N} \mathsf{E}_U(\mathfrak{A}_i\mathfrak{A}_j). \tag{22}$$

Let the *fundamental* state U_k correspond to the set of occupation numbers $a_r(U_k) = a_r$ $(r = 1, 2, \cdots)$. Then in the state U_k the system contains $\frac{1}{2}a_r(a_r - 1)$ pairs of particles with both members in the state $u_r(x)$ $(r = 1, 2, \cdots)$ and, for $r \neq s$, $a_r a_s$ pairs of particles with one member in the state $u_r(x)$ and the other in the state $u_s(x)$.

Now let λ_r and μ_r denote, respectively, the mathematical expectations of the quantities $\mathfrak{A}_i$ and $\mathfrak{A}_i^2$, when the ith particle is in the state $u_r(x)$ $(r = 1, 2, \cdots)$. These quantities do not depend on i and are determined by the structure of the particles and the choice of the quantities $\mathfrak{A}_i$. Formula (22) gives

$$\mathsf{E}_{U_k}(\mathfrak{A}^2) = \sum_{r=1}^{\infty} a_r\mu_r + 2\sum_{r=1}^{\infty} \tfrac{1}{2}a_r(a_r - 1)\lambda_r^2 + 2\sum_{\substack{r,s=1\\ r<s}}^{\infty} a_r a_s \lambda_r \lambda_s,$$

where it is understood that $a_r = a_r(U_k)$, $a_s = a_s(U_k)$. For the microcanonical average of the quantity we therefore find, by (17),

$$\begin{aligned} <\mathfrak{A}^2> &= (1/m) \sum_{k=1}^{m} \mathsf{E}_{U_k}\mathfrak{A}^2 \\ &= \sum_{r=1}^{\infty} <a_r>\mu_r + \sum_{r=1}^{\infty} <a_r^2>\lambda_r^2 - \sum_{r=1}^{\infty} <a_r>\lambda_r^2 \\ &\qquad + 2 \sum_{\substack{r,s=1\\ r<s}}^{\infty} <a_r a_s>\lambda_r\lambda_s \\ &= \sum_{r=1}^{\infty} (\mu_r - \lambda_r^2)<a_r> + \sum_{r=1}^{\infty} \sum_{s=1}^{\infty} \lambda_r\lambda_s<a_r a_s>. \end{aligned} \tag{23}$$

(In all of the above we assume, of course, that all the series obtained are absolutely convergent; for this, it suffices to assume that the numbers $|\lambda_r|$ and μ_r do not increase too rapidly with increasing r.) Since, on the other hand, by formula (16) of §5,

$$<\mathfrak{A}> = N\alpha = \sum_{r=1}^{\infty} \lambda_r<a_r>,$$

we have that

$$N^2\alpha^2 = \sum_{r=1}^{\infty} \sum_{s=1}^{\infty} \lambda_r\lambda_s<a_r><a_s>;$$

and formulas (23) and (21) give

$$\begin{aligned} D(\mathfrak{A}) &= \sum_{r=1}^{\infty} (\mu_r - \lambda_r^2)<a_r> \\ &\qquad + \sum_{r=1}^{\infty} \sum_{s=1}^{\infty} \lambda_r\lambda_s[<a_r a_s> - <a_r><a_s>]. \end{aligned} \tag{24}$$

This is the desired expression for the microcanonical dispersion of sum functions. To estimate the quantity $D(\mathfrak{A})$, we must know the mean values $<a_r>$ of the occupation numbers and the mean values $<a_r a_s>$ of their pairwise products. We have seen that the suitability of the mean value $<\mathfrak{A}>$ is guaranteed if the quantity

$$D(\mathfrak{A})/\delta\epsilon^2N^2$$

is sufficiently small. This requires that $D(\mathfrak{A})/N^2$ be negligibly small for small ϵ and δ. In Chapters IV and V we shall see that in actual physical cases, for $N \to \infty$, the quantity $D(\mathfrak{A})$ is infinitely large of order N. From this it follows that $D(\mathfrak{A})/N^2$ is of order $1/N$, i.e., that it actually becomes negligibly small.

We shall discuss in somewhat more detail the manner in which the suitability of mean values is connected with their experimental verification. Assume that the total energy of the system is E, but that no other information regarding its state is known. We then make a measurement of the quantity $\mathfrak{A}$ and want to know the probability

$$\mathsf{P}\{\mid \mathfrak{A} - \alpha N \mid < \epsilon N\}$$

that the value obtained for this quantity will lie between $(\alpha - \epsilon)N$ and $(\alpha + \epsilon)N$. However, our question is meaningless when phrased in this manner, because the theory gives the probability distribution of the quantity $\mathfrak{A}$ only if the state U, in which the system is found, is known. We know only its energy level E, which can correspond to an infinite set of different states U, and we have no way of knowing in which of these the system will be found. In order that our problem have a definite meaning, we must know the relative frequency with which the various states U of the manifold $\mathfrak{M}_E$ will occur, under the established conditions of the experiment, i.e., we must know the *probability law of the states* U, appropriate to the manner of experimentation. We suppose, for simplicity, that this law has a definite probability density $\rho(U)$. It is immediately clear that this probability density $\rho(U)$ may depend in an essential manner on the conditions of the experiment: for example, upon whether we measure the quantity $\mathfrak{A}$ a large number of times in one system with a definite interval of time between measurements, or in a large number of systems simultaneously. (The time interval in the former case must be sufficiently great so that the result of any measurement will not be influenced by the preceding one.) However, if the function $\rho(U)$ is known, then by the formula for total probability,

$$(25) \quad \mathsf{P}\{\mid \mathfrak{A} - \alpha N \mid > \epsilon N\} = \int_{\mathfrak{M}_E} \rho(U) \mathsf{P}_U\{\mid \mathfrak{A} - \alpha N \mid > \epsilon N\}\, d\omega,$$

where $d\omega$ is the volume element of the manifold $\mathfrak{M}_E$ for the measure we defined on this manifold in §2. Let M_δ denote, as before, the measure of the set of those states U for which

$$\mathsf{P}_U = \mathsf{P}_U\{\mid \mathfrak{A} - \alpha N \mid > \epsilon N\} > \delta.$$

Then, assuming that the function $\rho(U)$ is bounded on $\mathfrak{M}_E$, and denoting by R its upper bound, we conclude from (25) that

$$\mathsf{P}\{|\mathfrak{A} - \alpha N| > \epsilon N\} = \int_{\mathsf{P}_U \leq \delta} \rho(U)\mathsf{P}_U\, d\omega + \int_{\mathsf{P}_U > \delta} \rho(U)\mathsf{P}_U\, d\omega$$

$$\leq \delta \int_{\mathfrak{M}_E} \rho(U)\, d\omega + R \int_{\mathsf{P}_U > \delta} d\omega = \delta + RM_\delta;$$

whence, by (20),

$$\mathsf{P}\{|\mathfrak{A} - \alpha N| > \epsilon N\} \leq \delta + RD(\mathfrak{A})/\delta\epsilon^2 N^2. \tag{26}$$

If, as is the case in the great majority of physical problems, the quantity

$$D(\mathfrak{A})/N^2$$

approaches zero as $N \to \infty$, then for sufficiently small δ and ϵ, and sufficiently large N, the right side of (26) becomes arbitrarily small. This means that *for a sufficiently large number of particles, the approximate equality* $\mathfrak{A} \approx N\alpha$ *will agree with experiment with a relative error less than* $|\epsilon/\alpha|$ *with a probability arbitrarily close to unity.*

This conclusion, which gives a very satisfactory answer to the question of the real significance of our averages, is possible only if the probability distribution of the stationary states U has a bounded density $\rho(U)$. (In other respects this probability law can be arbitrary, and this constitutes the chief value of the results obtained.) By complicating the calculation somewhat, it could be shown that we arrive at the same result under the broader hypothesis that this law is absolutely continuous with respect to the measure introduced on the manifold $\mathfrak{M}_E$. However, if the probability law is not absolutely continuous with respect to this measure, then all of our calculations lose their basis, and the averages obtained by our method can lead to results which diverge sharply from experiment. We investigated a striking example of such a situation in §3 in considering the transition from complete statistics to symmetric (or antisymmetric) statistics. In general, the manifold $\mathfrak{S}_E$ of symmetric functions of the manifold $\mathfrak{M}_E$ has measure zero in the measure assigned by us (like a linear manifold of a smaller number of dimensions). Thus, if we consider a system obeying symmetric statistics, then the only states U which are really possible are those belonging to $\mathfrak{S}_E$, and hence

$$\int_{\mathfrak{S}_E} \rho(U)\, d\omega = 1.$$

This shows that the law $\rho(U)$ is not absolutely continuous with respect to our original measure; and, to make our conclusions agree with experiment, we had to replace the first method of averaging by a new one. This constituted the transition from complete statistics to symmetric (or antisymmetric) statistics.

Chapter IV

FOUNDATIONS OF THE STATISTICS OF PHOTONS

§1. Distinctive characteristics of the statistics of photons

In this chapter we shall study systems composed of photons (light quanta). Such systems possess certain distinctive properties which make them fundamentally different from systems composed of material particles. Even though all the methods we develop here are also applicable to systems of other types, the statistics of photons differ appreciably from the statistics of material particles and demand a separate study. We begin with the statistics of photons because they are considerably simpler than the statistics of material particles. Hence, the fundamental ideas of our method will not be burdened with too heavy a formal apparatus, and will emerge more clearly and be more easily understood. Then, after firmly grasping these ideas in a relatively simple mathematical context, the reader will be able to follow them without difficulty in the more complicated and formal relationships of the problems of the statistics of material particles.

The basic characteristic which distinguishes and simplifies the statistics of photons compared to the statistics of material particles is the fact that the number N of photons, comprising a system with a fixed total energy E, is not fixed, but changes from state to state. Thus, the set of all possible states of a system, for a given total energy E, is described by a set of eigenfunctions (belonging to the eigenvalue E), not of just one operator $\mathcal{H}$, but of the energy operators of all possible systems, composed of any number of photons.

On the other hand, *photons always obey symmetric statistics*. Therefore, if $\varepsilon_1, \varepsilon_2, \cdots, \varepsilon_r, \cdots$ is the sequence of possible photon energy levels for a given set of conditions (as usual, $0 \leq \varepsilon_1 \leq \varepsilon_2 \leq \cdots \leq \varepsilon_r \leq \cdots$, and each level is repeated a number of times equal to its multiplicity), then the number $\Omega(N, E)$ of linearly independent (symmetric) eigenfunctions of the operator $\mathcal{H}$, for a given level E and a given number of photons N, is equal to (III, §5) the number of solutions in terms of integral $a_r \geq 0$ of the system of equations

$$(1) \qquad \sum_{r=1}^{\infty} a_r = N, \qquad \sum_{r=1}^{\infty} a_r\varepsilon_r = E.$$

If we construct a linear basis of $\Omega(N, E)$ eigenfunctions for every N and then combine all these bases together, we get a system of

$$\Omega(E) = \sum_{N=1}^{\infty} \Omega(N, E)$$

(symmetric) eigenfunctions (of different operators), belonging to the same eigenvalue E. Each of these functions describes a possible state of our system when the system has a total energy E. Conversely, every such state is described by a linear combination of these functions. It is obvious that the number $\Omega(E)$ is simply equal to the number of solutions of the equation

$$\sum_{r=1}^{\infty} a_r \varepsilon_r = E$$

with integral $a_r \geq 0$, since the first of equations (1) does not apply when N is arbitrary. In analogy with III, §4, we shall call $\Omega(E)$ the *structure function* of our system of photons.

When the number N of particles is fixed, we saw (III, §2) that the microcanonical average of a phase function $f(U)$ would be the arithmetic average of its values for the $\Omega(N, E)$ functions of our linear basis. In other words, we agreed to use these $\Omega(N, E)$ functions with equal weights for averaging.

Now since the number N of photons has no fixed value, we naturally take as a basis for averaging, the system of $\Omega(E)$ functions just constructed. For the microcanonical average of a phase function $f(U)$ (for a given value E) we shall take the arithmetic average of its values for these $\Omega(E)$ functions. This principle of averaging is, of course, a new and arbitrary agreement: Functions which pertain to different N (and so depend on a different number of variables) were never equal to one another in statistical weight; now we ascribe the same weight to all of them, independent of the value of N. In §6 we shall strive to justify this new and arbitrary assumption by showing that for the cases of interest in statistical physics the results obtained by our method of averaging are independent of this method. Hence any other principle of averaging (within broad limits) would lead us to the same conclusions.

§2. Occupation numbers and their mean values

As we saw in III, §4, the symmetric eigenfunctions which constitute our linear basis can be chosen from the set of so-called "fundamental" functions of the system. In a state of the system described by such a function, the number a_r of particles in a state with energy level ε_r is fully determined. In other words, in each of the $\Omega(E)$ fundamental states comprising our chosen basis we have a completely determined set of "occupation numbers" a_r ($r = 1, 2, \cdots$) which are, of course, always subject to the requirement that

(2) $$\sum_{r=1}^{\infty} a_r \varepsilon_r = E.$$

The microcanonical averages of the occupation numbers a_r are therefore given by the formula

(3) $$\langle a_r \rangle = [\Omega(E)]^{-1} \sum_U a_r(U),$$

where the summation extends over all the fundamental functions U of our basis, and $a_r(U)$ denotes the quantity a_r in the state described by the function U.

A basic problem in the statistical thermodynamics of any system is to determine the distribution of the energy of the system among its component particles. In our theory we take the microcanonical average $<a_r>$ as the mean number of particles in an energy level ε_r. Therefore, the estimation of these quantities $<a_r>$ is a primary problem. Moreover, as we saw in III, §5, knowing the quantities $<a_r>$ allows us to write down immediately the microcanonical average of any sum function. It is chiefly by means of these sum functions that the state of a system is characterized in statistical physics. Further, as we saw in III, §6, to determine the suitability of the microcanonical averages (i.e., to justify the method of microcanonical averaging) in the case of sum functions it is also necessary to know the quantities $<a_r a_s>$ $(r, s = 1, 2, \cdots)$, which represent the microcanonical averages of the products of pairs of occupation numbers. Therefore, any mathematical apparatus which is employed in statistical physics must provide convenient analytical expressions for the quantities $<a_r>$ and $<a_r a_s>$. This statement pertains equally to particles of any type and to any of the three basic statistical schemes. In the following sections we shall look at the simplest example of photons to see how these expressions can be obtained.

Our first step will be to determine an elementary expression for the numbers $<a_r>$ and $<a_r a_s>$ in terms of the structure function $\Omega(E)$, which is the number of solutions of equation (2) with integral $a_r \geq 0$. For this purpose we note first of all that formula (3) can obviously be rewritten in the form

$$(4) \qquad <a_r> = [\Omega(E)]^{-1} \sum_{k=1}^{\infty} k \Lambda_k ,$$

where Λ_k denotes the number of those fundamental states (i.e., fundamental functions of our basis) in which $a_r = k$ $(k = 1, 2, \cdots)$. Now we denote by M_k $(k = 1, 2, \cdots)$ the number of those fundamental states in which $a_r \geq k$, so that

$$\Lambda_k = M_k - M_{k+1} \qquad (k = 1, 2, \cdots).$$

M_k is obviously equal to the number of those solutions of equation (2) in which $a_r \geq k$; or, equivalently, to the number of solutions in terms of integral $b_r \geq 0$ of the equation

$$b_1 \varepsilon_1 + \cdots + b_{r-1} \varepsilon_{r-1} + (k + b_r) \varepsilon_r + b_{r+1} \varepsilon_{r+1} + \cdots = E;$$

or

$$\sum_{i=1}^{\infty} b_i \varepsilon_i = E - k \varepsilon_r .$$

This means that $M_k = \Omega(E - k\varepsilon_r)$ $(k = 1, 2, \cdots)$, and hence that

$$\Lambda_k = \Omega(E - k\varepsilon_r) - \Omega[E - (k+1)\varepsilon_r] \qquad (k = 1, 2, \cdots).$$

Therefore, formula (4) gives

$$<a_r> = [\Omega(E)]^{-1}\sum_{k=1}^{\infty} k\{\Omega(E - k\varepsilon_r) - \Omega[E - (k+1)\varepsilon_r]\}.$$

By an elementary transformation due to Abel, we then find that

$$(5) \qquad <a_r> = \sum_{k=1}^{\infty} \Omega(E - k\varepsilon_r)/\Omega(E).$$

This simple formula will be the starting point of our further calculations.

We turn now to the derivation of an analogous expression for $<a_r a_s>$. We suppose first that $r \neq s$. Let Λ_{kl} denote the number of solutions of equation (2) in which $a_r = k$, $a_s = l$. Then, obviously,

$$(6) \qquad <a_r a_s> = [\Omega(E)]^{-1}\sum_{k=1}^{\infty}\sum_{l=1}^{\infty} kl\Lambda_{kl}.$$

On the other hand, if we denote by M_{kl} the number of solutions of (2) in which $a_r \geq k$, $a_s \geq l$, then, as is easily verified,

$$(7) \qquad \Lambda_{kl} = M_{kl} - M_{k+1,l} - M_{k,l+1} + M_{k+1,l+1}.$$

Further, in analogy with our previous calculation, we note that M_{kl} can be represented as the number of solutions in terms of integral $b_i \geq 0$ of the equation

$$b_1\varepsilon_1 + b_2\varepsilon_2 + \cdots + (k + b_r)\varepsilon_r + \cdots + (l + b_s)\varepsilon_s + \cdots = E;$$

or

$$\sum_{i=1}^{\infty} b_i\varepsilon_i = E - k\varepsilon_r - l\varepsilon_s.$$

Hence,

$$M_{kl} = \Omega(E - k\varepsilon_r - l\varepsilon_s).$$

Therefore, formulas (6) and (7) give

$$\begin{aligned}<a_r a_s> = [\Omega(E)]^{-1}\sum_{k=1}^{\infty}\sum_{l=1}^{\infty} kl\{\Omega(E - k\varepsilon_r - l\varepsilon_s) \\ - \Omega[E - (k+1)\varepsilon_r - l\varepsilon_s] - \Omega[E - k\varepsilon_r - (l+1)\varepsilon_s] \\ + \Omega[E - (k+1)\varepsilon_r - (l+1)\varepsilon_s]\};\end{aligned}$$

and, after twice applying the Abel transformation, we find

$$(8) \qquad <a_r a_s> = \sum_{k=1}^{\infty}\sum_{l=1}^{\infty} \Omega(E - k\varepsilon_r - l\varepsilon_s)/\Omega(E).$$

Finally, in the case $r = s$, the same argument applies in finding the required expression for the quantity $<a_r^2>$. Retaining the former notation, we have

$$\langle a_r^2 \rangle = [\Omega(E)]^{-1} \sum_{k=1}^{\infty} k^2 \Lambda_k = [\Omega(E)]^{-1} \sum_{k=1}^{\infty} k^2 (\mathrm{M}_k - \mathrm{M}_{k+1})$$
$$= [\Omega(E)]^{-1} \sum_{k=1}^{\infty} (2k - 1)\mathrm{M}_k ,$$

where the last expression is obtained from the penultimate one by the Abel transformation. Since we found that $\mathrm{M}_k = \Omega(E - k\varepsilon_r)$ $(k = 1, 2, \cdots)$, we obtain

$$(9) \qquad \langle a_r^2 \rangle = \sum_{k=1}^{\infty} (2k - 1)\Omega(E - k\varepsilon_r)/\Omega(E).$$

This completes the first stage of our investigation. Formulas (5), (8) and (9) give us expressions for the numbers $\langle a_r \rangle$, $\langle a_r a_s \rangle$ $(r \neq s)$ and $\langle a_r^2 \rangle$ in terms of the structure function Ω and these in turn reduce our problem to that of determining a convenient analytic expression for the structure function.

§3. Reduction to a problem of the theory of probability

The second step in our present investigation will be to reduce the problem of finding an asymptotic expression for the structure function Ω of our system, to a comparatively simple problem of the theory of probability. This step has great methodological significance because it establishes a bridge between the basic problems of statistical mechanics and the limit problems of the theory of probability. This bridge enables us to use the well-developed analytical methods of the modern theory of probability to solve problems in statistical physics.

However, before making this reduction, we must study in some detail the sequence

$$(10) \qquad \varepsilon_1 , \varepsilon_2 , \cdots , \varepsilon_r , \cdots$$

of possible energy levels of each of our photons. [In the sequence (10) each level is repeated a number of times equal to its multiplicity (degree of degeneracy).] First let our "photon gas" be enclosed in a vessel of unit volume. According to quantum physics, on the average, the number g_k of possible photon energy levels contained between k and $k + 1$ approaches infinity as $k \to \infty$. (We shall find the law which determines this increase in §5 of the present chapter.) In §2 of the Introduction, we agreed to approximate all the ε_r by integers. Therefore, we suppose that the integer k occurs g_k times in the sequence (10). This is the situation when the volume occupied by our system is unity. If this volume equals V (we always consider V to be a large integer), then, under very general conditions concerning the shape of the vessel, it can be shown from quantum physics that for sufficiently large k the number of possible energy levels in any (not too small) interval including k is increased by a factor V. We can assume this to be true for all k (small and large) without appreciably affecting the results in most physical problems. Therefore, we suppose that every integer k is repeated Vg_k times in the sequence (10). In particular, this means that the sequence (10) and the structure function $\Omega(E)$ depend strongly on the volume V occupied by our system. However, we must note that the asymptotic formulas which we shall derive in the following sections, are based on the assumption that E and V become infinitely large while maintaining some constant ratio (constant energy density E/V). Practically, this means that for the unit of energy we take approximately the average energy of an individual photon, so that E becomes a very large quantity (of the order of the average number of photons). For the volume we choose a unit so small that, on the average, there will be only a small number of photons per unit volume. The number V will then also be of the order of the average number of photons.

Finally, we make the following remark which will prove important later: Suppose we have any function $f(x)$ for which the series

$$\sum_{r=1}^{\infty} f(\varepsilon_r) \tag{11}$$

converges absolutely. Then from the preceding discussion it follows that

$$\sum_{r=1}^{\infty} f(\varepsilon_r) = V \sum_{k=1}^{\infty} g_k f(k). \tag{12}$$

Since the sum on the right side of this equation is independent of V, *every quantity of the form* (11) *in our asymptotic formulas must be thought of as infinitely large, proportional to* V.

Now we proceed to establish a probabilistic expression for the structure function Ω. Let β be a positive number, for the moment completely arbitrary. We consider random variables whose possible values are non-negative integral multiples of some fixed integer k, i.e., the numbers

$$0, k, 2k, \cdots, nk, \cdots.$$

Let the probability that the random variable assume the value nk be

$$e^{-\beta nk} \Big/ \sum_{n=0}^{\infty} e^{-\beta nk} = (1 - e^{-\beta k}) e^{-\beta nk} \quad (n = 0, 1, \cdots).$$

We denote this distribution by $p_k(x)$ and consider the sum of an infinite series of mutually independent random variables, of which g_k variables are distributed according to $p_k(x)$ $(k = 1, 2, \cdots)$. We shall show that this series converges with probability 1. For a variable distributed according to $p_k(x)$, the probability that it is different from zero is $e^{-\beta k}$. Therefore, the probability that at least one of the g_k terms of our series, which are distributed according to $p_k(x)$, is different from zero does not exceed $g_k e^{-\beta k}$. But the numbers g_k for most problems in quantum physics are such that the series

$$\sum_{k=1}^{\infty} g_k e^{-\beta k}$$

converges for all $\beta > 0$. Thus, with a probability as near as we please to unity, we can expect that *all* the terms of our series, beginning with the nth, will vanish if the number n is sufficiently large. But a series containing only a finite number of non-zero terms is trivially convergent. It therefore follows that the probability of convergence of our series is arbitrarily close to unity and hence equals 1, q.e.d.

The sum of our series is therefore a random variable whose distribution we denote by $P(x)$. Let S_V stand for the sum of V mutually independent random variables

$$S_V = \sum_{i=1}^{V} X_i ,$$

each of which has the distribution $P(x)$. We seek the distribution of the quantity S_V , i.e., the probability

$$\mathsf{P}(S_V = q)$$

that S_V is equal to a given non-negative integer q.

The quantity S_V is the sum of V mutually independent random variables X_i , each of which, being distributed according to $P(x)$, can in turn be considered as the sum of an infinite series of mutually independent random variables, among which there are g_k variables distributed according to $p_k(x)$. Thus we can consider the quantity S_V to be the sum of an infinite series of mutually independent random variables, among which there are Vg_k variables distributed according to $p_k(x)$. We shall denote these quantities by x_{kl} $(1 \leq l \leq Vg_k)$, so that

$$S_V = \sum_{k=1}^{\infty} \sum_{l=1}^{Vg_k} x_{kl} ,$$

where the x_{kl} obey the distribution $p_k(x)$, and all the x_{kl} are mutually independent. Hence, in order that $S_V = q$, it is necessary that

$$\sum_{k=1}^{\infty} \sum_{l=1}^{Vg_k} x_{kl} = q. \tag{13}$$

Since the random variable x_{kl} obeys the distribution $p_k(x)$, it can assume only values of the form kn_{kl} , where $n_{kl} \geq 0$ is an integer. Therefore, for the realization of equation (13), the probability of which we require, the system of values kn_{kl} assumed by the variables x_{kl} must be such that

$$\sum_{k=1}^{\infty} \sum_{l=1}^{Vg_k} kn_{kl} = q. \tag{K}$$

Because of the mutual independence of the variables x_{kl} the probability of their assuming the set of values $x_{kl} = kn_{kl}$ $(k = 1, 2, \cdots ; l = 1, 2, \cdots , Vg_k)$ is

$$\begin{aligned}\prod_{k=1}^{\infty} \prod_{l=1}^{Vg_k} \mathsf{P}(x_{kl} = kn_{kl}) &= \prod_{k=1}^{\infty} \prod_{l=1}^{Vg_k} (1 - e^{-\beta k}) e^{-\beta k n_{kl}} \\ &= \prod_{k=1}^{\infty} \{(1 - e^{-\beta k})^{Vg_k} \prod_{l=1}^{Vg_k} e^{-\beta k n_{kl}}\} \\ &= \{\prod_{k=1}^{\infty} (1 - e^{-\beta k})^{Vg_k}\} e^{-\beta Z},\end{aligned}$$

where $Z = \sum_{k=1}^{\infty} \sum_{l=1}^{Vg_k} kn_{kl}$. The probability of relation (13) is equal to a sum of probabilities of the form just written down, extended over all sets of values of the variables x_{kl} satisfying condition (K). This sum can be written as

$$\sum_{(K)} [\prod_{k=1}^{\infty} (1 - e^{-\beta k})^{Vg_k}]e^{-\beta Z} = [\prod_{k=1}^{\infty} (1 - e^{-\beta k})^{Vg_k}]e^{-\beta q} \sum_{(K)} 1.$$

The first factor

$$\prod_{k=1}^{\infty} (1 - e^{-\beta k})^{Vg_k}$$

is a completely determined function of the parameters β and V. Putting

$$\prod_{k=1}^{\infty} (1 - e^{-\beta k})^{-g_k} = \Phi(\beta),$$

we have

$$\prod_{k=1}^{\infty} (1 - e^{-\beta k})^{Vg_k} = \{\Phi(\beta)\}^{-V}.$$

The last factor $\sum_{(K)} 1$ is the number of solutions of equation (K) with integral $n_{kl} \geq 0$. It is easy to see that equation (K) differs only in notation from the equation

$$\sum_{r=1}^{\infty} a_r \varepsilon_r = q. \tag{14}$$

In fact, the number of levels ε_r equal to k in equation (14) is Vg_k, and the coefficients a_r of these levels can be denoted instead by n_{kl} $(1 \leq l \leq Vg_k)$. This then transforms equation (14) into equation (K). Thus,

$$\sum_{(K)} 1 = \Omega(q),$$

and we find

$$\mathsf{P}(S_V = q) = \{\Phi(\beta)\}^{-V} e^{-\beta q} \Omega(q),$$

whence

$$\Omega(q) = \{\Phi(\beta)\}^{V} e^{\beta q} \mathsf{P}(S_V = q). \tag{15}$$

This is the desired expression for the structure function. (It shows, in particular, that the possible values of the quantity S_V are the possible energy levels of the system. This could easily be seen directly from the definition of S_V.) We see that $\Omega(q)$ is very simply expressed in terms of the distribution $\mathsf{P}(S_V = q)$. Since we defined the random variable S_V as the sum of a very large number V of identically distributed and mutually independent random variables, we can use the highly developed analytical apparatus of the theory of probability to calculate approximately the probability $\mathsf{P}(S_V = q)$. Our problem is not complicated by the presence in formula (15) of the factor $\{\Phi(\beta)\}^{V}$ (which is independent of q), because formulas (5), (8) and (9) of the preceding section contain only ratios of structure functions of different arguments; hence this factor will cancel completely.

We note, finally, that the parameter β in formula (15) can have any positive value. We shall choose its value to give the asymptotic formulas the simplest possible form.

§4. Application of a limit theorem of the theory of probability

We now proceed to the third and final step in our study — the application of the local limit theorem of the theory of probability proved in I, §4.

It is first necessary to choose the value of the parameter β. To this end we prove the following auxiliary proposition:

LEMMA. *For any positive number C, the equation*

$$d \ln \Phi(\beta)/d\beta + C = 0$$

has a unique positive root.

Let us consider the function

$$\Phi_C(\beta) = e^{C\beta}\Phi(\beta).$$

It is obvious that $\Phi(\beta)$ and $\Phi_C(\beta)$ approach infinity as $\beta \to 0$. On the other hand, since $\Phi(\beta)$ is always greater than unity,

$$\Phi_C(\beta) > e^{C\beta}$$

and $\Phi_C(\beta)$ also approaches infinity as $\beta \to \infty$. Thus, the function $\Phi_C(\beta)$ and the function $\ln \Phi_C(\beta)$ approach infinity both for $\beta \to 0$ and for $\beta \to \infty$. However,

$$\ln \Phi_C(\beta) = C\beta + \ln \Phi(\beta),$$

and hence,

$$d^2 \ln \Phi_C(\beta)/d\beta^2 = d^2 \ln \Phi(\beta)/d\beta^2 = \textstyle\sum_{k=1}^{\infty} k^2 g_k e^{-\beta k}/(1 - e^{-\beta k})^2 > 0.$$

Therefore, the function $\ln \Phi_C(\beta)$ is convex on the entire half-line $(0, +\infty)$. Combining these properties, we see that

$$d \ln \Phi_C(\beta)/d\beta = C + d \ln \Phi(\beta)/d\beta$$

vanishes at precisely one point of the half-line. This proves the lemma.

Let our "photon gas" have energy E and volume V. Then we set β equal to the root (which exists and is unique by our lemma) of the equation

$$d \ln \Phi(\beta)/d\beta + E/V = 0. \tag{16}$$

Since our asymptotic formulas, as already noted, will be derived under the assumption that E and V are increasing without bound, but with the ratio E/V remaining constant, we can consider β to be constant in these formulas.

The quantity S_V, for whose distribution we seek an asymptotic expression, is the sum of V mutually independent random variables

$$X_i \qquad (1 \leq i \leq V),$$

each of which is distributed according to $P(x)$. This distribution (which, by the way, is independent of E and V) is the distribution of the sum

$$X_i = \sum_{k=1}^{\infty} \sum_{l=1}^{g_k} x_{kl},$$

where the variables x_{kl} are distributed according to $p_k(x)$. The mathematical expectation and the dispersion of the variable x_{kl} are, respectively, (as is easily verified)

$$a_{kl} = ke^{-\beta k}/(1 - e^{-\beta k})$$

and

$$b_{kl} = k^2 e^{-\beta k}/(1 - e^{-\beta k})^2.$$

It follows from a very elementary computation that the mathematical expectation and the dispersion of each of the variables X_i are, respectively,

$$a = \sum_{k=1}^{\infty} \sum_{l=1}^{g_k} a_{kl} = \sum_{k=1}^{\infty} kg_k e^{-\beta k}/(1 - e^{-\beta k})$$
$$= -\, d \ln \Phi(\beta)/d\beta$$

and

$$b = \sum_{k=1}^{\infty} \sum_{l=1}^{g_k} b_{kl} = \sum_{k=1}^{\infty} k^2 g_k e^{-\beta k}/(1 - e^{-\beta k})^2$$
$$= d^2 \ln \Phi(\beta)/d\beta^2.$$

For the mathematical expectation A and the dispersion B of the variable S_V, we therefore find the expressions

$$A = Va = -V\, d \ln \Phi/d\beta, \qquad B = Vb = V\, d^2 \ln \Phi/d\beta^2.$$

From relation (16) (i.e., from our choice for the value of the parameter β) we have, obviously, $A = E$. Therefore, the application of the one-dimensional local limit theorem in its simplified form [I, §4, (28)] gives

$$\text{(17)} \quad \mathsf{P}(S_V = E + u) = d(2\pi B)^{-\frac{1}{2}} e^{-u^2/2B} + O[V^{-\frac{3}{2}}(1 + |u|)],$$

where d denotes the spacing (in the sense defined in I, §2) of the sequence of energy levels of the system, and u is any integer such that $E + u$ is one of the levels.

Since

$$e^{-u^2/2B} = 1 + O(u^2/V),$$

it follows that

$$\mathsf{P}(S_V = E + u) = d(2\pi B)^{-\frac{1}{2}} + O[V^{-\frac{3}{2}}(1 + u^2)].$$

Putting $q = E + u$ in formula (15) we find

$$\Omega(E + u) = \{\Phi(\beta)\}^V e^{\beta(E+u)}\{d(2\pi B)^{-\frac{1}{2}} + O[V^{-\frac{3}{2}}(1 + u^2)]\};$$

and, in particular, for $u = 0$,

$$\Omega(E) = \{\Phi(\beta)\}^V e^{\beta E}\{d(2\pi B)^{-\frac{1}{2}} + O(V^{-\frac{3}{2}})\}.$$

Hence,

$$(18)\quad \begin{aligned} \Omega(E + u)/\Omega(E) &= e^{\beta u}\{1 + O[V^{-1}(1 + u^2)]\}\{1 + O(1/V)\}^{-1} \\ &= e^{\beta u}\{1 + O[V^{-1}(1 + u^2)]\}. \end{aligned}$$

Now we easily obtain a simple asymptotic expression for the mean values of the occupation numbers a_r. Using (18) and formula (5) of §2, we find

$$(19)\quad \begin{aligned} \langle a_r \rangle &= \sum_{m=1}^{\infty} e^{-\beta m \varepsilon_r}\{1 + O[V^{-1}(1 + m^2\varepsilon_r^2)]\} \\ &= \sum_{m=1}^{\infty} e^{-\beta m \varepsilon_r} + O(V^{-1}) = (e^{\beta\varepsilon_r} - 1)^{-1} + O(V^{-1}). \end{aligned}$$

Since the number k is repeated Vg_k times among the levels ε_r, the mean number of photons with energy k is

$$(20)\quad [Vg_k/(e^{\beta k} - 1)] + O(1).$$

Let $\mathfrak{A}$ be a sum function:

$$\mathfrak{A} = \sum_i \mathfrak{A}_i,$$

where the quantity $\mathfrak{A}_i$ depends only on the coordinates of the ith photon. The microcanonical average of the mathematical expectation of $\mathfrak{A}$ is, by III, §5, (16),

$$\langle \mathfrak{A} \rangle = \sum_{r=1}^{\infty} \langle a_r \rangle \lambda_r,$$

where λ_r is the mathematical expectation of $\mathfrak{A}_i$ (which was assumed to be the same for all i) in the photon state characterized by the energy level ε_r. (We recall that the quantity λ_r depends only on the form of the function $\mathfrak{A}_i$ and is determined independently of any statistical considerations.) The first of equations (19) gives us

$$\begin{aligned} \langle \mathfrak{A} \rangle &= \sum_{r=1}^{\infty} \lambda_r \sum_{m=1}^{\infty} e^{-\beta m \varepsilon_r}\{1 + O[V^{-1}(1 + m^2\varepsilon_r^2)]\} \\ &= \sum_{r=1}^{\infty} \lambda_r (e^{\beta\varepsilon_r} - 1)^{-1} + O\{V^{-1}[\sum_{r=1}^{\infty} |\lambda_r| (e^{\beta\varepsilon_r} - 1)^{-1} \\ &\qquad + \sum_{r=1}^{\infty} |\lambda_r| \varepsilon_r^2 \sum_{m=1}^{\infty} m^2 e^{-\beta m \varepsilon_r}]\}. \end{aligned}$$

If we suppose (as it actually happens in the great majority of real physical problems) that λ_r is a single-valued function of ε_r (i.e., that the quantity $\mathfrak{A}_i$ has the same mathematical expectation for all photon states which correspond to the same energy level), then all the series on the right side of the above equation have the form (11) (§3), and therefore by formula (12) are quantities proportional to V. (It is, of course, assumed that these series converge.) Hence we find

$$\begin{aligned} <\mathfrak{A}> &= \sum_{r=1}^{\infty} \lambda_r (e^{\beta\varepsilon_r} - 1)^{-1} + O(1) \\ &= V \sum_{k=1}^{\infty} g_k \Lambda_k (e^{\beta k} - 1)^{-1} + O(1), \end{aligned} \tag{21}$$

where $\Lambda_k = \lambda_r$ for $\varepsilon_r = k$. Thus, for a sum function the microcanonical average is always asymptotically proportional to V. This is, of course, immediately obvious. The asymptotic formula (21) defines the proportionality factor and shows that the relative error in the estimate obtained is extremely small.

In the case of photons, one of the most important sum functions is the number of photons N. For given E and V, this number can assume different values, depending on the state of the system (see §1). In other words, in a *definite* state of the system, the number N does not, in general, have a fixed value. Only its distribution is defined, so that it is only meaningful to speak of the microcanonical average $<N>$ *of the mathematical expectation* of the number N.

If $\mathfrak{A} = N$, then obviously $\mathfrak{A}_i = 1$, and hence $\lambda_r = 1$ $(r = 1, 2, \cdots)$. Therefore, we find

$$\begin{aligned} <N> &= \sum_{r=1}^{\infty} <a_r> = \sum_{r=1}^{\infty} (e^{\beta\varepsilon_r} - 1)^{-1} + O(1) \\ &= V \sum_{k=1}^{\infty} g_k (e^{\beta k} - 1)^{-1} + O(1). \end{aligned}$$

The mean number of photons per unit volume is thus

$$<N>/V = \sum_{k=1}^{\infty} g_k (e^{\beta k} - 1)^{-1} + O(V^{-1}).$$

§5. The Planck formula

To give a concrete meaning to the formulas derived in the preceding section, we must determine the numbers g_k and find the physical meaning of the parameter β. We shall consider the second problem in detail later, since extensive material, which we have not yet developed, is required to justify the universal value given to β. However, anticipating the result of a more detailed discussion, we simply say that in all cases we prescribe for the parameter β the value $1/kT$, where T is the absolute temperature of the system and k is a universal constant (the so-called Boltzmann constant). Thus, for any arbitrarily complicated system, composed of particles of any type, the parameter β always characterizes the temperature.

The determination of the numbers g_k , to which we now turn, can be accomplished in several physically different ways. One can start either from the wave theory of light, or partly from the wave theory and partly from the corpuscular theory, or, finally, from the purely corpuscular theory. Naturally, we choose the last way, since it is the only one which corresponds completely to the whole spirit of the theory we have developed. It must be pointed out, however, that the problem of the determination of the numbers g_k is in no way connected with statistical considerations. We solve this problem merely to present applications of our statistical conclusions which can be compared directly with experiment.

By definition, Vg_k denotes the number of linearly independent states of a photon in which its energy lies between k and $k + 1$. We must therefore write down the so-called "time-independent Schrödinger equation"

$$\mathcal{H}U = EU,$$

where $\mathcal{H}$ is the photon energy operator and E is a constant, and we must find the number of its linearly independent solutions for values of E between k and $k + 1$. However, the usual expression for the energy operator of a particle not under the influence of an external field,

$$\mathcal{H} = -(h^2/8\pi^2m^2)(\partial^2/\partial x^2 + \partial^2/\partial y^2 + \partial^2/\partial z^2),$$

cannot be employed in the case of photons, since $m = 0$ for a photon, and the connection between the energy and the momentum p is not given by the equation

$$\varepsilon = p^2/2m,$$

as it is for a (non-relativistic) material particle. [To put these results in their usual form, we use Planck's constant h directly, instead of $\hbar$ (Planck's constant divided by 2π). Both systems of notation are used in physics.] In order to determine the operator $\mathcal{H}$ correctly, we must start from the general relation between the energy and momentum of a particle, as given by the theory of relativity

$$\varepsilon = c(m^2c^2 + p^2)^{\frac{1}{2}},$$

where c denotes the velocity of light in vacuo. This relation is valid for particles of any type. In the case of photons, $m = 0$ and

$$\varepsilon = cp = c(p_x^2 + p_y^2 + p_z^2)^{\frac{1}{2}}.$$

(In the case of material particles if $p \ll mc$, then $\varepsilon \approx mc^2 + p^2/2m$.)

Thus, for photons p^2 is not proportional to ε, but to ε^2, and the operator of the quantity c^2p^2, having the form

$$-(c^2h^2/4\pi^2)(\partial^2/\partial x^2 + \partial^2/\partial y^2 + \partial^2/\partial z^2), \tag{22}$$

does not correspond to the energy, but to its square. We know (II, §3) that it follows from this that the eigenvalues of operator (22) are the squares of the eigenvalues of the operator $\mathcal{H}$, and the eigenfunctions of these two operators are the same. Hence, if we write the equation

$$-(c^2h^2/4\pi^2)(\partial^2U/\partial x^2 + \partial^2U/\partial y^2 + \partial^2U/\partial z^2) = E^2U,$$

or, equivalently,

$$\partial^2U/\partial x^2 + \partial^2U/\partial y^2 + \partial^2U/\partial z^2 = -4\pi^2E^2U/h^2c^2, \tag{23}$$

then its solutions will be the eigenfunctions of the photon energy operator, corresponding to the energy eigenvalue E. The potential energy of a photon, assuming that no external forces act on the system, is due only to the potential of the walls of the container which encloses our photon gas. Inside the container this potential is a constant which we may assume to be zero. It is well-known that a linear basis of the solutions of equation (23) can be obtained from expressions of the form

$$U = C \sin(ax + \eta) \sin(by + \zeta) \sin(cz + \xi), \tag{24}$$

where a, b, c, η, ζ, ξ, and C are constants, the first three of which are related by the equation

$$a^2 + b^2 + c^2 = 4\pi^2E^2/h^2c^2. \tag{25}$$

Now, since the probability of finding a photon in the neighborhood of the point (x, y, z) is proportional to $| U(x, y, z) |^2$ and must vanish outside the container (and hence, by continuity, also at the walls), we obtain well-defined boundary conditions, whose explicit formulation requires some assumption about the shape of the container. We assume that this container is the parallelopiped $0 \le x \le l_1$, $0 \le y \le l_2$, $0 \le z \le l_3$, with volume $l_1l_2l_3 = V$. Then if any of the six conditions $x = 0$, $y = 0$, $z = 0$, $x = l_1$, $y = l_2$, $z = l_3$ is satisfied, we must have $U = 0$. The first three conditions obviously require that $\eta = \zeta = \xi = 0$ in expression (24), so that we obtain

$$U = C \sin ax \sin by \sin cz, \qquad a^2 + b^2 + c^2 = 4\pi^2E^2/h^2c^2.$$

The requirement $U = 0$ for $x = l_1$, (independently of y and z) necessitates setting $a = n_1\pi/l_1$, where n_1 is an integer; the other two requirements lead to analogous relations, so that we must have

$$a = n_1\pi/l_1, \qquad b = n_2\pi/l_2, \qquad c = n_3\pi/l_3,$$

where n_1, n_2 and n_3 are integers which, by (25), satisfy the relation

$$n_1^2/l_1^2 + n_2^2/l_2^2 + n_3^2/l_3^2 = 4E^2/h^2c^2.$$

Thus, the possible values of the energy of the stationary states (i.e., the energy levels of a photon) are the numbers E, whose squares have the form

$$\tfrac{1}{4}h^2c^2(n_1^2/l_1^2 + n_2^2/l_2^2 + n_3^2/l_3^2),$$

where n_1, n_2 and n_3 are integers. Each triad of integers n_1, n_2, n_3 gives us one of the linearly independent eigenfunctions. More precisely, we must fix the signs of the numbers n_1, n_2 and n_3, since, for example, changing n_1 into $-n_1$ changes U into $-U$ and does not lead to an eigenfunction distinct from the preceding ones. For definiteness, therefore, we consider only the values $n_1 \geq 0$, $n_2 \geq 0$, $n_3 \geq 0$.

The number of linearly independent eigenfunctions corresponding to values of the energy $E \leq k$, where k is a given integer, is equal to the number of solutions with integral $n_1 \geq 0$, $n_2 \geq 0$, $n_3 \geq 0$ of the inequality

$$\tfrac{1}{4}h^2c^2(n_1^2/l_1^2 + n_2^2/l_2^2 + n_3^2/l_3^2) \leq k^2;$$

or, equivalently, to the number of points in the first octant, having integral coordinates and lying inside or on the boundary of the ellipsoid

$$\tfrac{1}{4}h^2c^2(x^2/l_1^2 + y^2/l_2^2 + z^2/l_3^2) = k^2. \tag{26}$$

This number is asymptotically (for large k) equal to one eighth of the volume of the ellipsoid (26), i.e., it is equal to

$$R(k) = \tfrac{4}{3}\pi(k^3/h^3c^3)l_1l_2l_3 = (4\pi V/3h^3c^3)k^3.$$

Hence, the number of linearly independent eigenfunctions belonging to energy levels between k and $k + dk$ is approximately

$$R'(k)\,dk = (4\pi V/h^3c^3)k^2\,dk,$$

where the interval dk must be sufficiently large so that the approximate formula has meaning, and at the same time small compared to k. However, in our previous notation this number is

$$V\textstyle\sum_{r=k}^{k+dk} g_r,$$

so that

$$\textstyle\sum_{r=k}^{k+dk} g_r = (4\pi/h^3c^3)k^2\,dk.$$

Thus, for $k < r < k + dk$, the value of the number g_r is, "on the average",

$$(4\pi/h^3c^3)k^2.$$

For the majority of calculations one can, without serious distortion of the results, assume directly that

$$g_k = (4\pi/h^3c^3)k^2, \tag{27}$$

at least for large values of k. However, one must recall that this is only an approximate calculation and that actually the number g_k depends on the arithmetic value of the number k in a more complicated way. In particular, there can be arbitrarily large numbers k which are not encountered among the energy levels of a photon, and for which therefore $g_k = 0$.

The number (27) is doubled by physicists because of quantum mechanical considerations which, from the point of view of the wave theory, correspond to the possibility of different polarizations of light. The fact is that the functions U, besides depending on their "classical" arguments x, y, z also depend on a peculiar quantum variable ("the spin"), which can assume only two values. Each of the eigenfunctions which we have defined up to now, therefore, describes not one but two different photon states, corresponding to the two values of the spin variable. Thus, finally, we obtain the approximate expression

$$g_k = (8\pi/h^3c^3)k^2,$$

which holds "on the average".

Substituting this expression in formula (20) of the preceding section, we find, for the average number of photons with energies between k and $k + 1$,

$$8\pi Vk^2/h^3c^3(e^{\beta k} - 1). \tag{28}$$

The energy ε of a photon is related to the frequency ν of the corresponding light wave by the universal equation $\varepsilon = h\nu$, where h is Planck's constant. The number of photons with frequencies between ν and $\nu + d\nu$ is therefore the number of photons with energies between $h\nu$ and $h\nu + h\, d\nu$, which, by formula (28) is, on the average,

$$[8\pi V(h\nu)^2/h^3c^3(e^{\beta h\nu} - 1)]h\, d\nu = 8\pi V\nu^2\, d\nu/c^3(e^{h\nu/kT} - 1)$$

(since $\beta = 1/kT$). Since the energy of every such photon is $h\nu$, the average energy assumed by the whole set of photons with frequencies lying between ν and $\nu + d\nu$ is

$$(8\pi hV/c^3)[\nu^3\, d\nu/(e^{h\nu/kT} - 1)],$$

and per unit volume

$$(8\pi h/c^3)[\nu^3\, d\nu/(e^{h\nu/kT} - 1)]. \tag{29}$$

This is the famous Planck formula, which has played an important role in the history of quantum physics. It gives the spectral distribution of energy of the so-called "black-body radiation". Fifty years ago, after many failures of the old classical theory, which constantly gave distributions for this spectrum in disagreement with experimental facts, Planck, as a result of hypotheses which were bold and risky for those times, arrived at his formula (29). It was found to be in excellent agreement with experiment. Planck's derivation of this formula, of course, has nothing in common with the one presented here, because Planck based his work on the tenets of the wave theory of light. (He could not have done otherwise in those times.) Planck could hardly have surmised that his bold idea would prove to be the "first swallow" of that tempestuous spring which in this half-century has radically changed the physics of elementary particles and has led mankind to great technical conquests, down to the realization of the ancient dream of mastering atomic energy.

§6. On the suitability of microcanonical averages

According to the general principles of statistical physics, the microcanonical averages of physical quantities are the values predicted by the statistical theory. It is these values which we compare with experimental data in order to check the theory. Experiment confirms the theory if the experimental value of a quantity is near its microcanonical average. However, if the possible values of the quantity are widely scattered, i.e., if in the majority of cases the values are markedly different from one another (and thus also markedly different from the microcanonical average), then we certainly cannot expect that a single experiment will give us a result which is near the calculated average. Further, situations are possible in which the experimental value will surely differ strongly from the average value (e.g., a game in which the player has probability $\frac{1}{2}$ of either winning or losing a large sum). In such cases, the average value (regardless of the method of averaging) can tell us nothing concerning the result of an experiment. We cannot even count on the average value of the results of a long series of experiments being near our microcanonical average: For this it would be necessary that the frequencies, with which the different possible states of the system occur in our series of experiments, conform to the weights given them in the microcanonical average. We have no reason to expect this. In the first place, we chose the microcanonical method of averaging for a number of theoretical reasons, and we did not take into account experimental conditions; but, it is obvious that the actual frequencies under discussion will always be different under different conditions of experimentation. Hence, no method of averaging can exist which would correspond in all cases to the actual experimental conditions.

The situation is entirely different if the quantity being studied is weakly dispersed, i.e., if the great majority of its possible values are not very different from one another. In this case, obviously, for any method of averaging (within broad limits) we shall find, for the average value, approximately the same number. This number will be close to the great majority of possible values of the quantity being studied. Hence, it will be close to the great majority of the results of our series of experiments. In this case we have every right to expect that a single experiment will give us a result near the computed average.

All these considerations were given, in substance, in Chapter III. We feel, however, that it is necessary to draw the attention of the reader to them repeatedly, because it is just these considerations which describe the connection between the physical world and the theory we have developed.

We saw in III, §6 that the "microcanonical dispersion"

$$D(\mathfrak{A}) = \langle \mathsf{E}[(\mathfrak{A} - \alpha)^2] \rangle$$

of a physical quantity $\mathfrak{A}$, with microcanonical average α, plays an important part in determining the suitability of this average. If, as the number of particles N approaches infinity (and the energy E of the system increases proportionately), the quantity $D(\mathfrak{A})$ becomes infinitely small compared to N^2 (or, equivalently, to E^2), then we can be sure that $|\mathfrak{A} - \alpha|$ will be as small as we like compared to $|\alpha|$ with a probability arbitrarily close to unity for any distribution (within wide limits) of frequencies of occurrence of the possible states of the system. This is precisely what we call the "suitability" of the microcanonical average α. This suitability will completely justify our arbitrarily chosen method of averaging and will permit us to expect close agreement between microcanonical averages and experimental facts. It will be guaranteed whenever we are able to prove the relation

$$D(\mathfrak{A}) = o(E^2).$$

We saw in the same section (III, §6) that if $\mathfrak{A}$ is a sum function, then $D(\mathfrak{A})$, in any of the three basic statistical schemes, is given by the expression

$$D(\mathfrak{A}) = \sum_{r=1}^{\infty} (\mu_r - \lambda_r^2) \langle a_r \rangle + \sum_{r=1}^{\infty} \sum_{s=1}^{\infty} \lambda_r \lambda_s [\langle a_r a_s \rangle - \langle a_r \rangle \langle a_s \rangle]. \tag{30}$$

Here it is assumed that

$$\mathfrak{A} = \sum_{i=1}^{N} \mathfrak{A}_i,$$

where the quantity $\mathfrak{A}_i$ depends only on the state of the ith particle. If the ith particle is in a stationary state with energy ε_r, then λ_r and μ_r denote the mathematical expectations of the quantities $\mathfrak{A}_i$ and $\mathfrak{A}_i^2$, respectively. (These expectations are assumed to be the same for all i.) To estimate the microcanonical dispersion $D(\mathfrak{A})$, we must find asymptotic expressions for the microcanonical averages $\langle a_r \rangle$ and $\langle a_r a_s \rangle$ of the occupation numbers and their pairwise products. We can foresee that we must obtain these asymptotic expressions rather accurately, since it is natural to expect that in the differences $\langle a_r a_s \rangle - \langle a_r \rangle \langle a_s \rangle$ a series of important terms will cancel one another. In fact, the accuracy to which we determined the numbers $<a_r>$ in §4, while fully satisfactory there, is quite insufficient for our present purpose. Therefore, the asymptotic calculation of these numbers must be carried out anew, on a more accurate basis. For this we need only a well-known form of the local limit theorem. The simplified formula (28) of I, §4 is no longer sufficient, and we must make use of the more accurate formula (27) of the same section. Instead of formula (17) of §4 we now obtain the more accurate formula

$$\mathsf{P}(S_V = E + u) = d(2\pi B)^{-\frac{1}{2}} e^{-u^2/2B} + (l_0 + l_1 u)B^{-\frac{3}{2}} + O[V^{-\frac{5}{2}}(V^{\frac{1}{2}} + |u|^3)],$$

where d and u have the same values as before, and l_0 and l_1 do not depend on either E or u. Using the fact that

$$e^{-u^2/2B} = 1 - \tfrac{1}{2}u^2 B^{-1} + O(u^4 B^{-2}),$$

we find

$$\mathsf{P}(S_V = E + u) = d(2\pi B)^{-\frac{1}{2}} + (l_0 + l_1 u + l_2 u^2)B^{-\frac{3}{2}} + O[V^{-\frac{5}{2}}(V^{\frac{1}{2}} + u^4)], \tag{31}$$

where $l_2 = -\frac{1}{2}d(2\pi)^{-\frac{1}{2}}$. With $q = E + u$, formula (15) of §3 therefore gives an expression for the structure function which is more accurate than the one we had in §4:

$$\Omega(E + u) = \{\Phi(\beta)\}^V e^{\beta(E+u)} \{d(2\pi B)^{-\frac{1}{2}} + (l_0 + l_1 u + l_2 u^2)B^{-\frac{3}{2}} + O[V^{-\frac{5}{2}}(V^{\frac{1}{2}} + u^4)]\};$$

and, in particular, for $u = 0$,

$$\Omega(E) = \{\Phi(\beta)\}^V e^{\beta E}\{d(2\pi B)^{-\frac{1}{2}} + l_0 B^{-\frac{3}{2}} + O(V^{-2})\}.$$

By an elementary calculation we obtain

$$\Omega(E + u)/\Omega(E) = e^{\beta u}\{1 + (k_1 u - \tfrac{1}{2}u^2)B^{-1} + O[V^{-2}(V^{\frac{1}{2}} + u^4)]\}, \tag{32}$$

where k_1 is a constant (independent of both E and u). This more accurate formula is used to replace formula (18) of §4. Now, proceeding to the derivation of the asymptotic formulas for $<a_r>$ and $<a_r a_s>$, we introduce the following convenient notation: We put

$$T_r = T_r(\beta) = (e^{\beta\varepsilon_r} - 1)^{-1} = \textstyle\sum_{m=1}^{\infty} e^{-\beta m \varepsilon_r}.$$

Then, clearly,

$$\varepsilon_r^k \textstyle\sum_{m=1}^{\infty} m^k e^{-\beta m \varepsilon_r} = (-1)^k\, d^k T_r/d\beta^k \equiv (-1)^k T_r^{(k)}.$$

In view of equation (32), formula (5) of §2 now gives us [in analogy with (19) of §4]

$$
\begin{aligned}
\langle a_r \rangle &= \textstyle\sum_{m=1}^{\infty} \Omega(E - m\varepsilon_r)/\Omega(E) \\
&= \textstyle\sum_{m=1}^{\infty} e^{-\beta m\varepsilon_r}\{1 - (k_1 m\varepsilon_r + \frac{1}{2}m^2\varepsilon_r^2)B^{-1} \\
&\qquad\qquad + O[V^{-2}(V^{\frac{1}{2}} + m^4\varepsilon_r^4)]\} \\
&= T_r + (k_1 T_r' - \tfrac{1}{2}T_r'')B^{-1} + O[V^{-2}(T_r V^{\frac{1}{2}} + T_r^{(4)})].
\end{aligned} \tag{33}
$$

This is a more accurate formula than equation (19) of §4. In our new notation the latter would have the simple form.

$$\langle a_r \rangle = T_r + O(V^{-1}).$$

Further, by using (32), for $r \neq s$, formula (8) of §2 gives

$$
\begin{aligned}
\langle a_r a_s \rangle &= \textstyle\sum_{k=1}^{\infty} \sum_{l=1}^{\infty} \Omega(E - k\varepsilon_r - l\varepsilon_s)/\Omega(E) \\
&= \textstyle\sum_{k=1}^{\infty} \sum_{l=1}^{\infty} e^{-\beta(k\varepsilon_r + l\varepsilon_s)}\{1 - B^{-1}[k_1(k\varepsilon_r + l\varepsilon_s) \\
&\quad + \tfrac{1}{2}(k\varepsilon_r + l\varepsilon_s)^2] + O[V^{-2}(V^{\frac{1}{2}} + (k\varepsilon_r + l\varepsilon_s)^4)]\} \\
&= T_r T_s + B^{-1}[k_1(T_r T_s' + T_s T_r') - \tfrac{1}{2}(T_r T_s'' + 2T_r' T_s' \\
&\quad + T_s T_r'')] + O[V^{-2}(V^{\frac{1}{2}} T_r T_s + T_r T_s^{(4)} + T_s T_r^{(4)})].
\end{aligned}
$$

At the same time formula (33) yields

$$
\begin{aligned}
\langle a_r \rangle \langle a_s \rangle = T_r T_s &+ B^{-1}[k_1(T_r T_s' + T_s T_r') - \tfrac{1}{2}(T_r T_s'' + T_s T_r'')] \\
&+ O[V^{-2}(V^{\frac{1}{2}} T_r T_s + T_r T_s^{(4)} + T_s T_r^{(4)})].
\end{aligned}
$$

From the last two formulas we find

$$
\begin{aligned}
\langle a_r a_s \rangle - \langle a_r \rangle \langle a_s \rangle &= -T_r' T_s' B^{-1} \\
&\quad + O[V^{-2}(V^{\frac{1}{2}} T_r T_s + T_r T_s^{(4)} + T_s T_r^{(4)})].
\end{aligned} \tag{34}
$$

Thus, the coefficient of microcanonical correlation of the numbers a_r and a_s ($r \neq s$), as was to be expected, is always negative and infinitely small in absolute magnitude, of order V^{-1}. This latter fact follows from equations (34) and (35) (see below).

Formula (34) provides the information necessary to estimate the microcanonical dispersion $D(\mathfrak{A})$. For this purpose we must first notice that in view of (33), the quantities $\langle a_r \rangle$ are asymptotically equal to the quantities T_r, which are functions of ε_r. Because of this, the first term on the right side of equation (30) is, by the general formula (12) of §3, asymptotically proportional to V. In the second (double) sum, we must distinguish between the terms where $r \neq s$ and those where $r = s$. In the terms of the first type, by (34), the numbers $\langle a_r a_s \rangle - \langle a_r \rangle \langle a_s \rangle$ are asymptotically equal to $-T_r' T_s' B^{-1}$. Hence, the double sum of these terms, on the basis of the same formula (12) of §3, gives a quantity which is asymptotically proportional to V^2B^{-1}, i.e., again proportional to V. Finally, the terms of the second type have the form $\lambda_r^2[\langle a_r^2\rangle - (\langle a_r\rangle)^2]$ and the sum over r extends from one to infinity. With the help of formulas (9) and (32), we can find the asymptotic expression for $\langle a_r^2\rangle - (\langle a_r\rangle)^2$:

$$(35) \qquad \langle a_r^2\rangle - (\langle a_r\rangle)^2 \approx T_r(T_r + 1).$$

Therefore, the sum of terms of the second type is asymptotically equal to

$$\sum_{r=1}^{\infty} \lambda_r^2 T_r(T_r + 1),$$

and hence, by formula (12) of §3, it is asymptotically proportional to V. (In all these estimates, of course, it is assumed that $|\lambda_r|$ and μ_r do not increase too rapidly with increasing r, so that all the series obtained are absolutely convergent.)

Formula (30) thus proves that $D(\mathfrak{A})$ is asymptotically proportional to V. But, as we saw above, in order to establish the suitability of the mean value $\langle\mathfrak{A}\rangle$, it is sufficient to have

$$D(\mathfrak{A}) = o(E^2);$$

or, equivalently,

$$D(\mathfrak{A}) = o(V^2).$$

Hence, the estimates we have obtained easily prove the suitability of the microcanonical averages of sum functions. As was noted above, this justifies, on the theoretical side, the arbitrary choice of the microcanonical method of averaging, and, on the practical side, explains the great value of the microcanonical averages for making comparisons with experimental results.

Chapter V

FOUNDATIONS OF THE STATISTICS OF MATERIAL PARTICLES

§1. Review of fundamental concepts

We shall now see that the methods developed in the preceding chapter can also be used to construct a statistical theory for a system of material particles. The basic new fact which must be considered in this transition is that the number N of particles composing the system is now rigidly fixed (just as the total energy E of the system). Hence, the possible states of a system, with a given total energy E and composed of N particles of a particular type, are now described by the eigenfunctions of the operator $\mathcal{H}$ which correspond to the eigenvalue E. In the case of complete statistics, all such eigenfunctions are "admissible"; but in the case of symmetric (or antisymmetric) statistics only symmetric (or antisymmetric) eigenfunctions are admissible. In all cases, we select a (finite) linear basis of mutually orthogonal and normalized functions from the family of all admissible eigenfunctions. The number of functions in this basis depends on N and E. We denote this number by $\Omega(N, E)$ and call it the structure function of the system. The fact that the structure function $\Omega(N, E)$ depends on two arguments is the reason that the statistics of material particles is more complicated in formal respects than the statistics of photons.

We saw in III, §4 that, in any of the three basic statistical schemes, the functions composing the above mentioned linear basis can be chosen from among the so-called "fundamental" eigenfunctions of the system. In a state described by such a function, the number a_r of particles in a state with energy level ε_r has a definite value. Hence, a uniquely defined set of "occupation numbers" a_r $(r = 1, 2, \cdots)$ corresponds to each "fundamental" eigenfunction which is a member of our basis. The numbers a_r must satisfy the relations

$$(K) \qquad \sum_{r=1}^{\infty} a_r = N, \qquad \sum_{r=1}^{\infty} a_r\varepsilon_r = E.$$

Conversely, a specific number of terms of the linear basis corresponds to each definite set of occupation numbers satisfying the equations (K). This number is different for the different statistical schemes. We saw (III, §5) that it can be represented as

$$(1) \qquad C(N)\prod_{r=1}^{\infty} \gamma(a_r),$$

where $C(N) = N!$ in the case of complete statistics, $C(N) = 1$ in the case of the other two statistics, and

$$\gamma(a) = 1/a! \qquad \text{(complete statistics)},$$

$$\gamma(a) = 1 \qquad \text{(symmetric statistics)},$$

$$\gamma(a) = \begin{cases} 1 & (a \leq 1) \\ 0 & (a > 1) \end{cases} \qquad \text{(antisymmetric statistics)}.$$

From the above we can obtain expressions for the structure function $\Omega(N, E)$ for all three cases. We find in the case of complete statistics

$$\Omega(N, E) = \sum_{(K)} N!\left(\prod_{r=1}^{\infty} a_r!\right)^{-1}, \tag{2}$$

in the case of symmetric statistics

$$\Omega(N, E) = \sum_{(K)} 1, \tag{3}$$

and in the case of antisymmetric statistics

$$\Omega(N, E) = \sum_{(K)}^{*} 1, \tag{4}$$

where in the first two cases the summation extends over all sets of integers $a_r \geq 0$ satisfying the conditions (K). The asterisk (*) in the third case indicates that the summation extends only over those solutions of the equations (K) for which $a_r \leq 1$ $(r = 1, 2, \cdots)$. These three equations can, of course, be combined into the single equation

$$\Omega(N, E) = C(N) \sum_{(K)} \prod_{r=1}^{\infty} \gamma(a_r),$$

where the summation extends over all sets of integers $a_r \geq 0$ $(r = 1, 2, \cdots)$ satisfying the relations (K).

As stated in III, §2, we take for the microcanonical average of any phase function (i.e., any quantity which assumes a definite value in each of the basic states of the system) the arithmetic mean of the values which it assumes in those $\Omega(N, E)$ states which are described by the eigenfunctions of the linear basis we selected. The justification for this choice of method of averaging can only be given later when the question of the suitability of microcanonical averages is considered.

The above statements constitute the starting points for the following investigations.

§2. Mean values of the occupation numbers

We have already had many opportunities to note that an important problem of a statistical theory is the determination of the mean values of the occupation numbers a_r as well as those of their pairwise products $a_r a_s$. In the photon case discussed in §2 of the preceding chapter, we expressed the mean values of the numbers a_r and $a_r a_s$ in terms of the structure function of the system. We must now take this first step for the case of material particles and, in addition, for all three basic statistical schemes.

Again, let $a_r(U)$ be the value of the number a_r in the state of the system described by the fundamental eigenfunction U. In the sequel we denote $1/\Omega(N, E)$ by Ω^{-1}. Then

$$(5) \qquad \langle a_r \rangle = \Omega^{-1} \sum_U a_r(U),$$

where the summation extends over all the fundamental functions of the linear basis.

1. *Complete Statistics*

In the case of complete statistics, to each set of occupation numbers a_r $(r = 1, 2, \cdots)$ satisfying the conditions (K) of §1 there correspond $N!\,(\prod_{r=1}^{\infty} a_r!)^{-1}$ different functions of the linear basis and, hence, this same number of terms is present in the sum on the right side of formula (5). Therefore, in the case of complete statistics

$$\langle a_r \rangle = \Omega^{-1} \sum_{(K)} N!\,(\prod_{i=1}^{\infty} a_i!)^{-1} a_r,$$

where the summation extends over all sets of occupation numbers which satisfy the conditions (K).

Since in the sum on the right side of this formula all the terms for which $a_r = 0$ vanish, the conditions (K) which have the form

$$a_i \geq 0 \qquad (i = 1, 2, \cdots), \qquad \sum_{i=1}^{\infty} a_i = N, \qquad \sum_{i=1}^{\infty} a_i \varepsilon_i = E,$$

can, without altering the result, be replaced by the conditions

$$(L) \quad a_r > 0, \qquad a_i \geq 0 \qquad (i \neq r), \qquad \sum_{i=1}^{\infty} a_i = N, \qquad \sum_{i=1}^{\infty} a_i \varepsilon_i = E.$$

Hence,

$$\langle a_r \rangle = N!\,\Omega^{-1} \sum_{(L)} a_r (\prod_{i=1}^{\infty} a_i!)^{-1},$$

where the summation extends over all systems of numbers a_i satisfying the conditions (L). Now, if we put $b_r = a_r - 1$, $b_i = a_i$ $(i \neq r)$, then the conditions (L) are equivalent to the conditions

$$(L') \quad \begin{array}{l} b_i \geq 0 \qquad (i = 1, 2, \cdots), \\ \qquad \sum_{i=1}^{\infty} b_i = N - 1, \qquad \sum_{i=1}^{\infty} b_i \varepsilon_i = E - \varepsilon_r, \end{array}$$

and we obtain

$$\langle a_r \rangle = N!\,\Omega^{-1} \sum_{(L')} (\prod_{i=1}^{\infty} b_i!)^{-1}.$$

In view of equation (2),

$$\sum_{(L')} (N - 1)!\,(\prod_{i=1}^{\infty} b_i!)^{-1} = \Omega(N - 1, E - \varepsilon_r),$$

and it then follows that

$$(6) \qquad \langle a_r \rangle = N \Omega^{-1} \Omega(N - 1, E - \varepsilon_r).$$

The mear values of the products $a_r a_s$ are found analogously. For $r \neq s$ we have

$$<a_r a_s> = N!\,\Omega^{-1} \sum_{(K)} a_r a_s \left(\prod_{i=1}^{\infty} a_i!\right)^{-1}.$$

The only terms of the sum $\sum_{(K)}$ different from zero are those in which $a_r > 0$, $a_s > 0$. Therefore,

$$<a_r a_s> = N!\,\Omega^{-1} \sum_{(K), a_r > 0, a_s > 0} a_r a_s \left(\prod_{i=1}^{\infty} a_i!\right)^{-1}.$$

Putting $b_r = a_r - 1$, $b_s = a_s - 1$, $b_i = a_i$ $(i \neq r, i \neq s)$, we obtain

$$<a_r a_s> = N!\,\Omega^{-1} \sum_{(M)} \left(\prod_{i=1}^{\infty} b_i!\right)^{-1},$$

where (M) denotes the set of conditions

$$(M) \qquad b_i \geq 0 \quad (i = 1, 2, \cdots), \qquad \sum_{i=1}^{\infty} b_i = N - 2, \qquad \sum_{i=1}^{\infty} b_i \varepsilon_i = E - \varepsilon_r - \varepsilon_s .$$

Again, in view of equation (2),

$$\sum_{(M)} (N-2)!\left(\prod_{i=1}^{\infty} b_i!\right)^{-1} = \Omega(N - 2, E - \varepsilon_r - \varepsilon_s),$$

and, therefore, we find

$$(7) \qquad <a_r a_s> = N(N-1)\Omega^{-1}\Omega(N - 2, E - \varepsilon_r - \varepsilon_s).$$

Finally, we find

$$\begin{aligned} <a_r(a_r - 1)> &= N!\,\Omega^{-1} \sum_{(K)} a_r(a_r - 1)\left(\prod_{i=1}^{\infty} a_i!\right)^{-1} \\ &= N!\,\Omega^{-1} \sum_{(K), a_r \geq 2} a_r(a_r - 1)\left(\prod_{i=1}^{\infty} a_i!\right)^{-1}. \end{aligned}$$

Setting $b_r = a_r - 2$, $b_i = a_i$ $(i \neq r)$ in the last expression it follows that

$$(8) \qquad \begin{aligned} <a_r(a_r - 1)> &= N!\,\Omega^{-1} \sum{}^{*} \left(\prod_{i=1}^{\infty} b_i!\right)^{-1} \\ &= N(N-1)\Omega^{-1}\Omega(N - 2, E - 2\varepsilon_r), \end{aligned}$$

where $\sum^*$ denotes summation over all b_i such that $b_i \geq 0$, $\sum_{i=1}^{\infty} b_i = N - 2$, and $\sum_{i=1}^{\infty} b_i \varepsilon_i = E - 2\varepsilon_r$. Therefore, in virtue of equation (6)

$$(9) \qquad \begin{aligned} <a_r^2> &= <a_r(a_r - 1)> + <a_r> \\ &= \Omega^{-1}\{N(N-1)\Omega(N-2, E-2\varepsilon_r) + N\Omega(N-1, E-\varepsilon_r)\}. \end{aligned}$$

2. *Symmetric Statistics*

In the case of symmetric statistics formula (3) of §1 shows that $\Omega(N, E)$ is the number of solutions of the system of equations

$$\sum_{i=1}^{\infty} a_i = N, \qquad \sum_{i=1}^{\infty} a_i \varepsilon_i = E$$

with integral $a_i \geq 0$. We denote by Γ_k $(k = 1, 2, \cdots)$ the number of solutions of the same system which satisfy the subsidiary condition $a_r \geq k$. Then, evidently, $\Gamma_k - \Gamma_{k+1}$ will be the number of solutions of the system of equations

$$a_r = k, \quad a_i \geq 0 \quad (i = 1, 2, \cdots),$$
$$\sum_{i=1}^{\infty} a_i = N, \quad \sum_{i=1}^{\infty} a_i \varepsilon_i = E,$$

whence

$$\langle a_r \rangle = \Omega^{-1} \sum_{k=1}^{\infty} k(\Gamma_k - \Gamma_{k+1}).$$

Using the Abel transformation, we easily obtain from this the result

$$\langle a_r \rangle = \Omega^{-1} \sum_{k=1}^{\infty} \Gamma_k .$$

But, on the other hand, putting $b_r = a_r - k$, $b_i = a_i$ $(i \neq r)$, we find that Γ_k is the number of solutions, with integral $b_i \geq 0$, of the system

$$\sum_{i=1}^{\infty} b_i = N - k, \quad \sum_{i=1}^{\infty} b_i \varepsilon_i = E - k\varepsilon_r ,$$

i.e.,

$$\Gamma_k = \Omega(N - k, E - k\varepsilon_r),$$

and, therefore,

$$(10) \qquad \langle a_r \rangle = \Omega^{-1} \sum_{k=1}^{\infty} \Omega(N - k, E - k\varepsilon_r).$$

Furthermore, for $r \neq s$, we denote by Γ_{kl} $(k = 1, 2, \cdots ; l = 1, 2, \cdots)$ the number of solutions of the system

$$\sum_{i=1}^{\infty} a_i = N, \quad \sum_{i=1}^{\infty} a_i \varepsilon_i = E \qquad (a_i \geq 0, a_r \geq k, a_s \geq l).$$

Then, as is easily calculated, the number of solutions of the system

$$\sum_{i=1}^{\infty} a_i = N, \quad \sum_{i=1}^{\infty} a_i \varepsilon_i = E \qquad (a_i \geq 0, a_r = k, a_s = l)$$

is equal to

$$\Gamma_{kl} - \Gamma_{k+1,l} - \Gamma_{k,l+1} + \Gamma_{k+1,l+1} .$$

Hence,

$$\langle a_r a_s \rangle = \Omega^{-1} \sum_{k,l=1}^{\infty} kl\{\Gamma_{kl} - \Gamma_{k+1,l} - \Gamma_{k,l+1} + \Gamma_{k+1,l+1}\},$$

and, by using the Abel transformation twice,

$$\langle a_r a_s \rangle = \Omega^{-1} \sum_{k,l=1}^{\infty} \Gamma_{kl} .$$

Putting

$$b_i = a_i \quad (i \neq r, i \neq s), \quad b_r = a_r - k, \quad b_s = a_s - l,$$

we easily find as before that

$$\Gamma_{kl} = \Omega(N - k - l, E - k\varepsilon_r - l\varepsilon_s).$$

Consequently,

$$(11) \qquad <a_r a_s> = \Omega^{-1}\textstyle\sum_{k,l=1}^{\infty} \Omega(N - k - l, E - k\varepsilon_r - l\varepsilon_s).$$

Finally, a completely analogous computation yields

$$<a_r^2> = \Omega^{-1}\textstyle\sum_{k=1}^{\infty} k^2(\Gamma_k - \Gamma_{k+1}) = \Omega^{-1}\sum_{k=1}^{\infty} (2k - 1)\Gamma_k ,$$

from which it follows that

$$(12) \qquad <a_r^2> = \Omega^{-1}\textstyle\sum_{k=1}^{\infty} (2k - 1)\Omega(N - k, E - k\varepsilon_r).$$

3. *Antisymmetric Statistics*

In this case, the numbers a_r can assume only the values 0 and 1. If we denote by $\Omega_r(N, E)$ the number of those solutions of the system

$$(N) \qquad \textstyle\sum_{i=1}^{\infty} a_i = N, \qquad \sum_{i=1}^{\infty} a_i\varepsilon_i = E, \qquad 0 \leq a_i \leq 1,$$

in which $a_r = 1$, then obviously

$$<a_r> = \Omega^{-1}\Omega_r(N, E).$$

Denoting by $\Omega_{r'}(N, E)$, the number of those solutions of the systen (N) in which $a_r = 0$, we have on the one hand that

$$\Omega_r(N, E) + \Omega_{r'}(N, E) = \Omega(N, E).$$

But, on the other hand, putting $b_i = a_i$ $(i \neq r)$, $b_r = a_r - 1$, we see that $\Omega_r(N, E)$ equals the number of those solutions of the system

$$\textstyle\sum_{i=1}^{\infty} b_i = N - 1, \qquad \sum_{i=1}^{\infty} b_i\varepsilon_i = E - \varepsilon_r , \quad 0 \leq b_i \leq 1,$$

in which $b_r = 0$, i.e., $\Omega_r(N, E)$ equals $\Omega_{r'}(N - 1, E - \varepsilon_r)$. Consequently,

$$\Omega_r(N, E) = \Omega(N - 1, E - \varepsilon_r) - \Omega_r(N - 1, E - \varepsilon_r).$$

Successive application of this recurrence formula yields

$$\Omega_r(N, E) = \textstyle\sum_{k=1}^{\infty} (-1)^{k-1}\Omega(N - k, E - k\varepsilon_r),$$

and, therefore,

$$(13) \qquad <a_r> = \Omega^{-1}\textstyle\sum_{k=1}^{\infty} (-1)^{k-1}\Omega(N - k, E - k\varepsilon_r).$$

Now, suppose $r \neq s$ and let 1) $\Omega_{rs}(N, E)$, 2) $\Omega_{rs'}(N, E)$, 3) $\Omega_{r's}(N, E)$ and 4) $\Omega_{r's'}(N, E)$ denote, respectively, the numbers of those solutions of the system (N) in which 1) $a_r = a_s = 1$, 2) $a_r = 1$, $a_s = 0$, 3) $a_r = 0$, $a_s = 1$, and 4) $a_r = a_s = 0$. Then, first of all

$$\Omega(N, E) = \Omega_{rs}(N, E) + \Omega_{rs'}(N, E) + \Omega_{r's}(N, E) + \Omega_{r's'}(N, E),$$

and

$$<a_r a_s> = \Omega^{-1}\Omega_{rs}(N, E).$$

On the other hand, our customary method of transforming from the numbers a_i to the numbers b_i easily yields

$$\begin{aligned}\Omega_{rs}(N, E) &= \Omega_{r's}(N-1, E-\varepsilon_r) = \Omega_{rs'}(N-1, E-\varepsilon_s) \\ &= \Omega_{r's'}(N-2, E-\varepsilon_r-\varepsilon_s) \\ &= \Omega(N-2, E-\varepsilon_r-\varepsilon_s) - \Omega_{rs}(N-2, E-\varepsilon_r-\varepsilon_s) \\ &\quad - \Omega_{r's}(N-2, E-\varepsilon_r-\varepsilon_s) - \Omega_{rs'}(N-2, E-\varepsilon_r-\varepsilon_s).\end{aligned}$$

By repeated use of this formula, we find

$$\Omega_{rs}(N, E) = \sum_{k,l=1}^{\infty} (-1)^{k+l}\Omega(N-k-l, E-k\varepsilon_r-l\varepsilon_s),$$

and, therefore (for $r \neq s$),

$$(14) \quad \langle a_r a_s \rangle = \Omega^{-1}\sum_{k,l=1}^{\infty} (-1)^{k+l}\Omega(N-k-l, E-k\varepsilon_r-l\varepsilon_s).$$

Finally, in the case of antisymmetric statistics it is always true that $a_r^2 = a_r$, so that in virtue of (13)

$$(15) \quad \langle a_r^2 \rangle = \langle a_r \rangle = \Omega^{-1}\sum_{k=1}^{\infty} (-1)^{k-1}\Omega(N-k, E-k\varepsilon_r).$$

By looking at the set of formulas (6)–(15) and recalling the definition of Ω^{-1}, we see now that in all three statistical schemes the mean values $\langle a_r \rangle$, $\langle a_r a_s \rangle$ $(r \neq s)$ and $\langle a_r^2 \rangle$ are expressed very simply by ratios of the form

$$(16) \quad \Omega(N-u, E-v)/\Omega(N, E),$$

where u and v are positive numbers. Consequently, if we find sufficiently simple and accurate estimates of (16), then we shall be able to obtain the required asymptotic estimates for the mean values of the occupation numbers, their squares and their pairwise products. We shall find such estimates in the next section.

However, we shall first derive one more simple formula which will be needed later. We know that in the sequence of energy levels ε_r (of the individual particles comprising the system) the same number can occur more than once, i.e., it is possible that $\varepsilon_r = \varepsilon_s$ for $r \neq s$. In this case, for brevity, let us put $\langle a_r a_s \rangle = \tau_r$. We obtain an expression for the quantity τ_r for our three basic statistical schemes by putting $\varepsilon_s = \varepsilon_r$ in formulas (7), (11) and (14).

In the case of complete statistics formulas (7) and (8) easily yield

$$(17) \quad \langle a_r^2 \rangle - \langle a_r \rangle = \langle a_r(a_r-1) \rangle = \tau_r.$$

In the case of symmetric statistics we find, in virtue of (10), (12) and (11),

$$(18) \quad \begin{aligned}\langle a_r^2 \rangle - \langle a_r \rangle &= 2\Omega^{-1}\sum_{m=1}^{\infty} (m-1)\Omega(N-m, E-m\varepsilon_r) \\ &= 2\Omega^{-1}\sum_{k,l=1}^{\infty} \Omega(N-k-l, E-k\varepsilon_r-l\varepsilon_r) \\ &= 2\tau_r.\end{aligned}$$

Finally, in the case of antisymmetric statistics $a_r^2 = a_r$, and consequently,

$$<a_r^2> - <a_r> = 0. \tag{19}$$

By using the "index of symmetry" σ of a system, introduced in III, §3, formulas (17), (18) and (19) can be combined into the one formula

$$<a_r^2> - <a_r> = (\sigma + 1)\tau_r, \tag{20}$$

which, therefore, holds for all three fundamental statistical schemes.

§3. Reduction to a problem of the theory of probability

Now we must consider the reduction of the problem of finding an asymptotic estimate for the function $\Omega(N, E)$ to a certain limit problem of the theory of probability. We indicated the general method by which this can be done in §3 of the previous chapter. However, we must take into account two essential facts which differentiate our new problem from the one we solved in IV, §3: 1) We dealt only with symmetric statistics in the photon case, whereas now we must include all three schemes; 2) The structure function $\Omega(E)$ in the photon case depended only on the single variable E; now $\Omega(N, E)$ depends on the variables N and E in each of the three fundamental schemes. This last fact implies that the solution of our new problem will require use of the two-dimensional limit theorems of the theory of probability.

As in the previous chapter, let us denote by g_k the number of energy levels of an individual particle included between k and $k + 1$, when the system occupies unit volume. Following the same reasoning used in IV, §3, we again conclude that the number g_k becomes Vg_k when the system occupies the volume V. Hence, the structure function $\Omega(N, E)$ depends strongly on V. In deriving our asymptotic formulas, we will assume that the numbers N, E and V approach infinity but remain in constant ratios. Hence, equation (12) of §3 of the preceding chapter remains valid so that here also the sum of each absolutely convergent series of the form

$$\sum_{i=1}^{\infty} f(\varepsilon_i)$$

is a quantity proportional to V (and to each of the numbers N, E).

Let us now introduce the two-dimensional distribution $p_k(x, y)$. This distribution is defined for each integer k as follows:

1) The random variable x can assume only non-negative integral values;
2) If $x = n$ $(n = 0, 1, \cdots)$, then $y = nk$;
3) The probability that $x = n$, $y = nk$, is

$$\begin{aligned} &\mathsf{P}(x = n, y = nk) \\ &\quad = \gamma(n)e^{-(\alpha+\beta k)n}\left[\sum_{i=0}^{\infty} \gamma(i)e^{-(\alpha+\beta k)i}\right]^{-1} \qquad (n = 0, 1, \cdots), \end{aligned} \tag{21}$$

where α and β are parameters, whose values we shall choose below, and $\gamma(n)$ is a function of the integral, non-negative variable n. We introduced $\gamma(n)$ in Chapter III and defined it for the various kinds of statistics as follows:

$$(\mathrm{I})\quad \begin{cases} 1)\ \text{for complete statistics,}\quad \gamma(n) = 1/n!, \\ 2)\ \text{for symmetric statistics,}\quad \gamma(n) = 1, \\ 3)\ \text{for antisymmetric statistics} \\ \qquad \gamma(n) = \begin{cases} 1 & \text{for } n = 0 \text{ and } n = 1, \\ 0 & \text{for } n > 1. \end{cases} \end{cases}$$

It is understood that the values of the parameters α and β must be chosen so that the series in brackets on the right side of (21) converges for all $k \geq 1$.

The distribution $p_k(x, y)$ just defined is evidently *degenerate*: All pairs of possible values of (x, y) obeying this law are located on one half of the line $y = kx$. However, this fact will not be important in our development.

Now let us consider the infinite sequence (x_i, y_i) $(i = 1, 2, \cdots)$ of pairs of random variables among which there are

$$g_1 \text{ pairs subject to the law } p_1(x, y),$$

$$g_2 \text{ pairs subject to the law } p_2(x, y)$$

and, in general,

$$g_k \text{ pairs subject to the law } p_k(x, y) \qquad (k = 1, 2, \cdots).$$

We assume that the pairs with different subscripts are mutually independent.

The probability that the pair of random variables (x, y), distributed according to the law $p_k(x, y)$, have a value different from zero can be found from equation (21). This probability is

$$\sum_{n=1}^{\infty} \gamma(n) e^{-(\alpha+\beta k) n} \left[\sum_{i=0}^{\infty} \gamma(i) e^{-(\alpha+\beta k) i}\right]^{-1}$$
$$= e^{-(\alpha+\beta k)} \sum_{n=0}^{\infty} \gamma(n+1) e^{-(\alpha+\beta k) n} \left[\sum_{i=0}^{\infty} \gamma(i) e^{-(\alpha+\beta k) i}\right]^{-1} \leq e^{-\alpha-\beta k},$$

since $\gamma(n + 1) \leq \gamma(n)$ $(n = 0, 1, 2, \cdots)$ for all three kinds of statistics. Consequently, the probability that at least one of the g_k pairs of our sequence, distributed according to the law $p_k(x, y)$, have a value different from zero is less than

$$g_k e^{-\alpha-\beta k}.$$

In the case of material particles, as we shall see below, the quantum theory gives different values for the numbers g_k from those found in the case of photons. However, here too the series

$$\sum_{k=1}^{\infty} g_k e^{-\beta k}$$

always converges for all $\beta > 0$. As in the photon case, we therefore easily conclude that the series

$$\sum_{i=1}^{\infty} x_i = X, \qquad \sum_{i=1}^{\infty} y_i = Y$$

both converge with probability 1. The distribution of the pair of random variables (X, Y) is clearly not degenerate. It will be denoted by $P(X, Y)$. Evidently the form of this law depends only on the parameters α and β and on the nature of the particles composing the system. The law $P(X, Y)$ plays a fundamental role in the remainder of this chapter.

As in the photon case, we consider the volume V occupied by our system to be an integer. Let us consider V mutually independent pairs of random variables (X_i, Y_i) $(i = 1, 2, \cdots, V)$ distributed according to the same law $P(X, Y)$. Let us put

$$\sum_{i=1}^{V} X_i = S_V, \qquad \sum_{i=1}^{V} Y_i = T_V,$$

$$C(p) = p! \qquad \text{for complete statistics,}$$

$$C(p) = 1 \qquad \text{for the other two statistics.}$$

Let us also assume that the series

$$\sum_{p=0}^{\infty} \sum_{q=0}^{\infty} e^{-\alpha p - \beta q} \Omega(p, q)/C(p) = \Phi(\alpha, \beta)$$

converges for the values of the parameters α and β which will be chosen below. Then we obtain the following theorem:

THEOREM. *For any pair of non-negative integers p, q,*

$$(22) \qquad \Omega(p, q) = C(p)\Phi(\alpha, \beta)e^{\alpha p + \beta q}\mathsf{P}(S_V = p, T_V = q),$$

where the last factor on the right side denotes the probability of the simultaneous fulfillment of the equalities $S_V = p$, $T_V = q$.

Proof. For brevity, let us put

$$\{\sum_{n=0}^{\infty} \gamma(n)e^{-nz}\}^{-1} = \Gamma(z).$$

Then $p_k(x, y)$ can be written more briefly as

$$\mathsf{P}(x = n, y = nk) = \Gamma(\alpha + \beta k)\gamma(n)e^{-n(\alpha+\beta k)} \qquad (n = 0, 1, \cdots).$$

We defined the pair of random variables (S_V, T_V) as the sum of V mutually independent pairs (X_i, Y_i) $(i = 1, 2, \cdots, V)$ each of which is distributed according to the law $P(X, Y)$. This law $P(X, Y)$ is, in turn, the distribution of the sum of an infinite series of mutually independent random pairs among which there are g_k pairs distributed according to the law $p_k(x, y)$ $(k = 1, 2, \cdots)$. This shows that the pair (S_V, T_V) can be considered as the sum of an infinite series of mutually independent pairs among which there are Vg_k pairs distributed according to the law

$$p_k(x, y) \qquad (k = 1, 2, \cdots).$$

We will denote these latter random pairs by (x_{kl}, y_{kl}) $(l = 1, 2, \cdots, Vg_k)$, so that

$$S_V = \sum_{k=1}^{\infty} \sum_{l=1}^{Vg_k} x_{kl}, \qquad T_V = \sum_{k=1}^{\infty} \sum_{l=1}^{Vg_k} y_{kl}.$$

The pair (x_{kl}, y_{kl}) is, of course, distributed according to the law $p_k(x, y)$.

For the variables S_V, T_V to have the values p, q, respectively, it is necessary that the values $x_{kl} = n_{kl}$, $y_{kl} = kn_{kl}$, assumed by the random variables x_{kl}, y_{kl}, should satisfy the relations

$$(P) \qquad \sum_{k=1}^{\infty} \sum_{l=1}^{Vg_k} n_{kl} = p, \qquad \sum_{k=1}^{\infty} k \sum_{l=1}^{Vg_k} n_{kl} = q.$$

Let n_{kl} $(1 \leq k < \infty, 1 \leq l \leq Vg_k)$ be a definite set of non-negative integers satisfying the equations (P). Then, because of the mutual independence of the pairs (x_{kl}, y_{kl}), the probability that $x_{kl} = n_{kl}$, $y_{kl} = kn_{kl}$ $(1 \leq k < \infty, 1 \leq l \leq Vg_k)$ will be

$$
\begin{aligned}
&\prod_{k=1}^{\infty} \prod_{l=1}^{Vg_k} \mathsf{P}(x_{kl} = n_{kl}, y_{kl} = kn_{kl}) \\
&\quad = \prod_{k=1}^{\infty} \prod_{l=1}^{Vg_k} \Gamma(\alpha + \beta k)\gamma(n_{kl})e^{-n_{kl}(\alpha+\beta k)} \\
(23) \qquad &\quad = \{\prod_{k=1}^{\infty} [\Gamma(\alpha + \beta k)]^{Vg_k}\}e^{-\alpha Z-\beta Z'} \prod_{k=1}^{\infty} \prod_{l=1}^{Vg_k} \gamma(n_{kl}) \\
&\quad = e^{-\alpha p-\beta q}\{\prod_{k=1}^{\infty} [\Gamma(\alpha + \beta k)]^{Vg_k}\}\prod_{k=1}^{\infty} \prod_{l=1}^{Vg_k} \gamma(n_{kl}),
\end{aligned}
$$

where $Z = \sum_{k=1}^{\infty} \sum_{l=1}^{Vg_k} n_{kl}$, $Z' = \sum_{k=1}^{\infty} k \sum_{l=1}^{Vg_k} n_{kl}$.

This is the probability that the random variables x_{kl}, y_{kl}, respectively, will assume definite values n_{kl}, kn_{kl} which satisfy the equations (P). Because of the above, the probability that $S_V = p$, $T_V = q$ is the sum of probabilities of the type (23) extended over the whole system of non-negative integers n_{kl} which satisfy equations (P), i.e.,

$$
\begin{aligned}
&\mathsf{P}(S_V = p, T_V = q) \\
(24) \qquad &\quad = e^{-\alpha p-\beta q}\{\prod_{k=1}^{\infty} [\Gamma(\alpha + \beta k)]^{Vg_k}\} \sum_{(P)} \prod_{k=1}^{\infty} \prod_{l=1}^{Vg_k} \gamma(n_{kl}).
\end{aligned}
$$

Now, if we recall that in the sequence of energy levels

$$\varepsilon_1, \varepsilon_2, \cdots, \varepsilon_r, \cdots$$

of a particle the integer k appears Vg_k times, then we note that the sum

$$\sum_{(P)} \prod_{k=1}^{\infty} \prod_{l=1}^{Vg_k} \gamma(n_{kl})$$

on the right side of (24) differs only in notation from the sum

$$\sum_{(K)} \prod_{r=1}^{\infty} \gamma(a_r),$$

which is used (see III, §5) to define $\Omega(p, q)$, if the conditions (K) have the form

$$(K) \qquad \sum_{r=1}^{\infty} a_r = p, \qquad \sum_{r=1}^{\infty} a_r \varepsilon_r = q.$$

[In the case of complete statistics, the sum $\sum_{(K)}$ must be multiplied by $p!$ in order to obtain $\Omega(p, q)$.] Therefore, we obtain

$$C(p) \sum_{(P)} \prod_{k=1}^{\infty} \prod_{l=1}^{Vg_k} \gamma(n_{kl}) = \Omega(p, q),$$

where $C(p) = p!$ for complete statistics and $C(p) = 1$ for the other two statistics. Hence equation (24) yields

$$(25) \quad \mathsf{P}(S_V = p, T_V = q) = e^{-\alpha p - \beta q} \{\prod_{k=1}^{\infty} [\Gamma(a + \beta k)]^{Vg_k}\} \Omega(p, q)/C(p).$$

Summing this relation over p and q from 0 to ∞, we find

$$1 = \{\prod_{k=1}^{\infty} [\Gamma(\alpha + \beta k)]^{Vg_k}\} \sum_{p,q=0}^{\infty} \{e^{-\alpha p - \beta q} \Omega(p, q)/C(p)\}$$
$$= \{\prod_{k=1}^{\infty} [\Gamma(\alpha + \beta k)]^{Vg_k}\} \Phi(\alpha, \beta).$$

Hence,

$$(26) \qquad \prod_{k=1}^{\infty} [\Gamma(\alpha + \beta k)]^{Vg_k} = \{\Phi(\alpha, \beta)\}^{-1},$$

and therefore (25) yields

$$\mathsf{P}(S_V = p, T_V = q) = e^{-\alpha p - \beta q} \Omega(p, q)/\Phi(\alpha, \beta) C(p),$$

which is equivalent to (22). Thus, our theorem is proved.

Formula (22) reduces the study of the properties of the function $\Omega(p, q)$ to the investigation of the distribution of the pair of random variables (S_V, T_V). This pair represents the sum of an infinitely large number V of mutually independent random pairs which are identically distributed. Hence, we are led to one of the most thoroughly discussed limit problems of the theory of probability for which a very accurate solution is known. Let us note that the presence of the factor $\Phi(\alpha, \beta)$ on the right side of (22) cannot cause any difficulty because the expressions for the mean values of the various phase functions always contain only *ratios* of structure functions Ω. Consequently, this factor always cancels in such expressions.

Let us make another remark of subsequent interest. In addition to depending on the parameters α and β, the function $\Phi(\alpha, \beta)$ depends on the form of the function $\Omega(p, q)$ which in turn depends on the volume V occupied by the system. Hence, the function $\Phi(\alpha, \beta)$ also depends on V. The form of this dependence is very simple as equation (26) shows: Denoting by $\Phi_1(\alpha, \beta)$ the expression for the function $\Phi(\alpha, \beta)$ when $V = 1$, we find

$$(27)\quad \Phi(\alpha, \beta) = \{\Phi_1(\alpha, \beta)\}^V, \qquad \ln \Phi(\alpha, \beta) = V \ln \Phi_1(\alpha, \beta).$$

Thus, the function $\ln \Phi(\alpha, \beta)$ is directly proportional to the volume V occupied by the system. It is easy to see that the dependence of the structure function $\Omega(p, q)$ on the volume has a considerably more complex character.

§4. Choice of values for the parameters α and β

The values of the parameters α and β were until now restricted only by the general requirement that the series of interest to us converge. Otherwise, these values remained arbitrary. Now, before proceeding to the application of the limit theorem of the theory of probability, it will be expedient for us to select these values so that subsequent computations will be as simple as possible. The present section is devoted to this selection.

In §3, we defined the function

$$\Phi(\alpha, \beta) = \sum_{p,q=0}^{\infty} e^{-\alpha p - \beta q}\Omega(p, q)/C(p)$$

for all three statistical schemes. Further, we showed [formula (26)] that

$$\Phi(\alpha, \beta) = \prod_{k=1}^{\infty} [\Gamma(\alpha + \beta k)]^{-Vg_k};$$

or, equivalently, that

$$(28)\quad \Phi(\alpha, \beta) = \prod_{r=1}^{\infty} [\Gamma(\alpha + \beta\varepsilon_r)]^{-1} = \prod_{r=1}^{\infty} \{\sum_{n=0}^{\infty} \gamma(n) e^{-n(\alpha+\beta\varepsilon_r)}\},$$

where the function $\gamma(n)$ for each of the three statistical schemes is defined according to rule (I) of §3.

Let us denote by E_0 the smallest possible value of the energy of a system composed of N particles of a given type. It is obvious that in the case of complete or symmetric statistics $E_0 = N\varepsilon_1$. However, in the case of antisymmetric statistics, where not more than one particle can be found in the state characterized by the level ε_r, we will have $E_0 = \sum_{r=1}^{N} \varepsilon_r$.

Now let us prove the following general proposition:

THEOREM: *Let a system composed of N particles have energy $E > E_0$. Then the set of equations*

$$(29)\qquad \partial \ln \Phi/\partial\alpha = -N, \qquad \partial \ln \Phi/\partial\beta = -E$$

has a unique solution (α, β) for $\beta > 0$.

Proof. We consider the function

$$F(\alpha, \beta) = e^{\alpha N + \beta E}\Phi(\alpha, \beta) = e^{\alpha N+\beta E}\prod_{r=1}^{\infty} \{\sum_{n=0}^{\infty} \gamma(n) e^{-n(\alpha+\beta\varepsilon_r)}\},$$

and we study its behavior in the half-plane $\beta > 0$. First, we establish successively that

1°. $F(\alpha, \beta) \to \infty$ uniformly with respect to α $(-\infty < \alpha < \infty)$ as $\beta \to 0$.
2°. $F(\alpha, \beta) \to \infty$ uniformly with respect to α $(-\infty < \alpha < \infty)$ as $\beta \to \infty$.
3°. $F(\alpha, \beta) \to \infty$ uniformly with respect to β $(0 < \beta < \beta_0)$ as $|\alpha| \to \infty$.
(β_0 is any positive number.)

In order to prove 1°, we note that $\gamma(0) = \gamma(1) = 1$ for all statistics and, therefore,

$$F(\alpha, \beta) \geq e^{\alpha N + \beta E} \prod_{r=1}^{\infty} \{1 + e^{-\alpha - \beta \varepsilon_r}\}.$$

Let A be an arbitrarily large integer and let β be small enough so that $e^{-\beta \varepsilon_r} > \frac{1}{2}$ $(r \leq A)$. Then

$$\ln F(\alpha, \beta) > \alpha N + \sum_{r=1}^{A} \ln \{1 + e^{-\alpha - \beta \varepsilon_r}\} > \alpha N + A \ln \{1 + \tfrac{1}{2} e^{-\alpha}\} \equiv f(\alpha).$$

An elementary computation easily shows that the function $f(\alpha)$ has its smallest value for $\alpha = \ln [(A - N)/2N] \equiv \alpha_0$. Consequently,

$$\ln F(\alpha, \beta) > f(\alpha) \geq f(\alpha_0) > N\alpha_0 = N \ln [(A - N)/2N].$$

Since A can be as large as desired for small enough β, statement 1° is proved.

Let us turn to the proof of 2° and 3°. First let us assume that we are concerned only with complete or symmetric statistics, so that $E_0 = N\varepsilon_1$, and $\gamma(n) > 0$ for any $n \geq 0$. Since $\gamma(0) = 1$,

$$\sum_{k=0}^{\infty} \gamma(k) e^{-k(\alpha + \beta \varepsilon_r)} > 1, \qquad r \geq 1,$$

and in virtue of (28),

$$\Phi(\alpha, \beta) \geq \sum_{k=0}^{\infty} \gamma(k) e^{-k(\alpha + \beta \varepsilon_1)}.$$

Similarly for any $m \geq 0$

$$\Phi(\alpha, \beta) \geq \gamma(m) e^{-m(\alpha + \beta \varepsilon_1)},$$

whence

$$F(\alpha, \beta) \geq \gamma(m) e^{\alpha(N-m) + \beta(E - m\varepsilon_1)}. \tag{30}$$

Putting $m = N$, we find

$$F(\alpha, \beta) \geq \gamma(N) e^{\beta(E - E_0)}.$$

Since $E > E_0$ by hypothesis, 2° is proved. Further, putting $m = N + 1$ and $m = N - 1$ in (30), we find

$$F(\alpha, \beta) \geq \gamma(N + 1) e^{-\alpha - \beta \varepsilon_1},$$

and

$$F(\alpha, \beta) \geq \gamma(N - 1) e^{\alpha + \beta \varepsilon_1}.$$

The first of these inequalities shows that $F(\alpha, \beta) \to \infty$ uniformly as

$\alpha \to -\infty$, $(0 < \beta < \beta_0)$. The second shows that $F(\alpha, \beta) \to \infty$ uniformly as $\alpha \to \infty$, $(\beta > 0)$. Hence, statement 3° is also proved.

Now, let us turn to the case of antisymmetric statistics. Here $E_0 = \sum_{r=1}^{N} \varepsilon_r$ and

$$(31) \quad \begin{aligned} F(\alpha, \beta) &= e^{\alpha N + \beta E} \prod_{r=1}^{\infty} (1 + e^{-\alpha - \beta \varepsilon_r}) \\ &= e^{\alpha N + \beta E} \{\prod_{r=1}^{N} (e^{\alpha + \beta \varepsilon_r} + 1) e^{-(\alpha + \beta \varepsilon_r)}\} \prod_{r=N+1}^{\infty} (1 + e^{-\alpha - \beta \varepsilon_r}). \end{aligned}$$

Therefore,

$$F(\alpha, \beta) > e^{\alpha N + \beta E} e^{-(\alpha N + \beta E_0)} \prod_{r=1}^{N} (1 + e^{\alpha + \beta \varepsilon_r}) = e^{\beta(E - E_0)} \prod_{r=1}^{N} (1 + e^{\alpha + \beta \varepsilon_r}).$$

Because $E > E_0$, this inequality proves 2°. But here we evidently also obtain 3° for the case $\alpha \to \infty$. In order to prove 3° for the case $\alpha \to -\infty$, it is sufficient to note that, just as before, it follows from (31) that

$$F(\alpha, \beta) > e^{\beta(E - E_0)} (1 + e^{-\alpha - \beta \varepsilon_{N+1}}).$$

From this we find that $F(\alpha, \beta) \to \infty$ uniformly as $\alpha \to -\infty$ in any interval $0 < \beta < \beta_0$. Hence, statements 1°, 2° and 3° are proved for all three kinds of statistics.

It evidently follows from the three statements proved above that $F(\alpha, \beta)$ approaches infinity uniformly as $\beta \to 0$ and as $\alpha^2 + \beta^2 \to \infty$. Therefore, it assumes a smallest value at a certain point (α, β) in the half-plane $\beta > 0$. At this point we have

$$\partial \ln F(\alpha, \beta)/\partial \alpha = N + \partial \ln \Phi(\alpha, \beta)/\partial \alpha = 0,$$

$$\partial \ln F(\alpha, \beta)/\partial \beta = E + \partial \ln \Phi(\alpha, \beta)/\partial \beta = 0$$

and, hence, that

$$\partial \ln \Phi/\partial \alpha = -N, \qquad \partial \ln \Phi/\partial \beta = -E.$$

Thus, the existence of a solution of the system of equations (29) in the half-plane $\beta > 0$ has been proved; the uniqueness still remains to be proved. For this purpose, let us assume that the point (α', β'), which is different from the point (α, β), also satisfies this system of equations. Then, since the second derivatives of the function $\ln F$ coincide with the corresponding derivatives of the function $\ln \Phi$,

$$\begin{aligned} &\ln F(\alpha, \beta) - \ln F(\alpha', \beta') \\ &\quad = \tfrac{1}{2}\{(\alpha - \alpha')^2 \partial^2 \ln \Phi/\partial \alpha^2 + 2(\alpha - \alpha')(\beta - \beta') \partial^2 \ln \Phi/\partial \alpha \partial \beta \\ &\qquad + (\beta - \beta')^2 \partial^2 \ln \Phi/\partial \beta^2\}, \end{aligned}$$

where all three second order derivatives are evaluated at the same point

$$[\alpha' + \theta(\alpha - \alpha'), \beta' + \theta(\beta - \beta')] \qquad (0 < \theta < 1).$$

We quickly arrive at the desired contradiction if we show that the quadratic form on the right side of the last equality is positive definite, because then

$$\ln F(\alpha, \beta) - \ln F(\alpha', \beta') > 0,$$

which is impossible in view of the definition of the point (α, β). Thus, we must still prove that

$$\partial^2 \ln \Phi/\partial\alpha^2 > 0,$$
$$(\partial^2 \ln \Phi/\partial\alpha^2)(\partial^2 \ln \Phi/\partial\beta^2) - (\partial^2 \ln \Phi/\partial\alpha\partial\beta)^2 > 0$$

identically in the region $\beta > 0$.

We have

$$\ln \Phi(\alpha, \beta) = \sum_{r=1}^{\infty} \ln \{\sum_{n=0}^{\infty} \gamma(n) e^{-n(\alpha+\beta\varepsilon_r)}\}$$
$$= \sum_{r=1}^{\infty} \ln S_r(\alpha, \beta),$$

where

$$S_r(\alpha, \beta) = \sum_{n=0}^{\infty} \gamma(n) e^{-n(\alpha+\beta\varepsilon_r)}.$$

Hence,

$$(32) \quad \partial^2 \ln \Phi/\partial\alpha^2 = \sum_{r=1}^{\infty} S_r^{-2}\{S_r\partial^2 S_r/\partial\alpha^2 - (\partial S_r/\partial\alpha)^2\} = \sum_{r=1}^{\infty} T_r,$$

where

$$T_r = S_r^{-2}\{S_r\partial^2 S_r/\partial\alpha^2 - (\partial S_r/\partial\alpha)^2\}.$$

Since

$$\partial S_r/\partial\alpha = -\sum_{n=0}^{\infty} n\gamma(n) e^{-n(\alpha+\beta\varepsilon_r)},$$

and

$$\partial^2 S_r/\partial\alpha^2 = \sum_{n=0}^{\infty} n^2\gamma(n) e^{-n(\alpha+\beta\varepsilon_r)},$$

the Schwarz inequality easily yields

$$T_r > 0 \qquad (r = 1, 2, \cdots),$$

and therefore, by (32),

$$\partial^2 \ln \Phi/\partial\alpha^2 > 0.$$

Furthermore, it is easy to see that if a differentiation with respect to α is replaced by a differentiation with respect to β in any partial derivative of the sum S_r, then the partial derivative is multiplied by ε_r. Therefore, it follows from (32) that

$$\partial^2 \ln \Phi/\partial\alpha\partial\beta = \sum_{r=1}^{\infty} \varepsilon_r T_r,$$
$$\partial^2 \ln \Phi/\partial\beta^2 = \sum_{r=1}^{\infty} \varepsilon_r^2 T_r.$$

Hence, in the half-plane $\beta > 0$, in virtue of the Schwarz inequality, we conclude that

$$(33) \qquad (\partial^2 \ln \Phi/\partial\alpha^2)(\partial^2 \ln \Phi/\partial\beta^2) - (\partial^2 \ln \Phi/\partial\alpha\partial\beta)^2 > 0.$$

This proves our theorem.

In all that follows, we will understand α and β to be numbers satisfying the set of equations (29). The existence and uniqueness of this pair of numbers have just been proved.

Let us make another important remark. Because of the relations (27) in §3, the set of equations (29), which defines the values of the parameters α and β, is equivalent to the set

$$(34) \qquad \partial \ln \Phi_1/\partial\alpha = -N/V, \qquad \partial \ln \Phi_1/\partial\beta = -E/V.$$

As was stated at the start of §3, we shall subsequently let the numbers N, E and V become infinitely large while holding their ratios constant. Hence, the right sides of (34) will be constants. Since the form of the function $\Phi_1(\alpha, \beta)$ is independent of N, V and E, the selected values of the parameters α and β are also independent of them. Hence, we will always consider the numbers α and β as constants.

§5. Application of a limit theorem of the theory of probability

Now we shall establish asymptotic formulas for the distribution of the random pair (S_V, T_V) which was constructed in §3. We stated at the end of §3 that this problem reduces to one of the most thoroughly discussed limit problems of the theory of probability, since the conditions necessary to apply the central limit theorem (in its local form) are satisfied and, moreover, we have the simplest case of identically distributed and mutually independent terms.

We must first find the mathematical expectations of the quantities S_V and T_V. Since

$$\Phi(\alpha, \beta) = \sum_{pq,=0}^{\infty} e^{-\alpha p-\beta q}\Omega(p, q)/C(p),$$

then, by (22) and (29),

$$\begin{aligned} \mathsf{E}S_V &= \sum_{p,q=0}^{\infty} p\, \mathsf{P}(S_V = p, T_V = q) \\ &= [\Phi(\alpha, \beta)]^{-1} \sum_{p,q=0}^{\infty} pe^{-\alpha p-\beta q}\Omega(p, q)/C(p) \\ &= -[\Phi(\alpha, \beta)]^{-1}[\partial\Phi(\alpha, \beta)/\partial\alpha] = -\partial \ln \Phi(\alpha, \beta)/\partial\alpha \\ &= N. \end{aligned}$$

Similarly,

$$\mathsf{E}T_V = -\partial \ln \Phi(\alpha, \beta)/\partial\beta = E.$$

Thus, the mathematical expectations of the quantities S_V and T_V equal

N and E, respectively, for the values of the parameters α and β which we selected. These results will impart especial simplicity to our asymptotic formulas.

Now we must determine the second central moments of the quantities S_V and T_V. We find

$$\begin{aligned} B_{11} &= \mathsf{E}\{[S_V - \mathsf{E}S_V]^2\} = \mathsf{E}\{S_V^2\} - \{(1/\Phi)(\partial\Phi/\partial\alpha)\}^2 \\ &= \sum_{p,q=0}^{\infty} p^2\, \mathsf{P}(S_V = p, T_V = q) - \{(1/\Phi)(\partial\Phi/\partial\alpha)\}^2 \\ &= (1/\Phi)\sum_{p,q=0}^{\infty} p^2 e^{-\alpha p - \beta q}\Omega(p, q)/C(p) - \{(1/\Phi)(\partial\Phi/\partial\alpha)\}^2 \\ &= (1/\Phi)(\partial^2\Phi/\partial\alpha^2) - \{(1/\Phi)(\partial\Phi/\partial\alpha)\}^2 \\ &= \partial^2 \ln \Phi(\alpha, \beta)/\partial\alpha^2, \end{aligned}$$

and therefore, by (27) (§3),

$$B_{11} = V\partial^2 \ln \Phi_1(\alpha, \beta)/\partial\alpha^2 = Vb_{11},$$

where $b_{11} = \partial^2 \ln \Phi_1/\partial\alpha^2$ is a constant.

In exactly the same way, we find

$$B_{12} = \mathsf{E}\{S_V T_V - \mathsf{E}S_V \mathsf{E}T_V\} = \partial^2 \ln \Phi(\alpha, \beta)/\partial\alpha\partial\beta = Vb_{12},$$

$$B_{22} = \mathsf{E}\{[T_V - \mathsf{E}T_V]^2\} = \partial^2 \ln \Phi(\alpha, \beta)/\partial\beta^2 = Vb_{22},$$

where b_{12} and b_{22} are constants. Here, in virtue of (33) of §4,

$$B_{11}B_{22} - B_{12}^2 = V^2(b_{11}b_{22} - b_{12}^2) = V^2\Delta > 0.$$

To estimate the probability $\mathsf{P}(S_V = N + u_1, T_V = E + u_2)$ we use Theorem 2 of Chapter I. The vectors (X_i, Y_i), which compose the vector (S_V, T_V), are non-degenerate mutually independent integral-valued random vectors (§3), identically distributed according to the law $P(X, Y)$. In view of I, §2, the vector (X_i, Y_i) has the maximal lattice $a_0 + k\alpha + l\beta$, $b_0 + k\gamma + l\delta$, where $|\alpha\delta - \beta\gamma| = |d| > 0$.* Thus, the possible values of the vector (S_V, T_V) belong to the lattice $Va_0 + k\alpha + l\beta$, $Vb_0 + k\gamma + l\delta$. From equation (22) of §3 it follows that these possible values are simultaneously those values of the arguments of the structure function $\Omega(p, q)$ for which Ω is not zero. In other words, all the physically possible pairs of values of the number of particles and the total energy of the system are included in this lattice. Therefore, we must put

$$N + u_1 = Va_0 + k\alpha + l\beta, \qquad E + u_2 = Vb_0 + k\gamma + l\delta.$$

Since N and E are, respectively, the mathematical expectations of the

* The two different meanings of each of the symbols α, β can be clearly distinguished by context.

quantities S_V and T_V, it follows that u_1 and u_2, in formulas (38) and (39) of Theorem 2 of Chapter I, have the same values as here. Moreover, we must evidently write V instead of n. Hence, formula (39) of Chapter I gives

$$\begin{aligned} &\mathsf{P}(S_V = N + u_1, T_V = E + u_2) \\ &\quad = |d| (2\pi V\Delta^{\frac{1}{2}})^{-1} e^{-(1/2\Delta V)(b_{11}u_2^2 - 2b_{12}u_1u_2 + b_{22}u_1^2)} + O[V^{-2}(1 + u)], \end{aligned} \tag{35}$$

where we put $u = |u_1| + |u_2|$. Similarly, formula (38) of Chapter I yields the more accurate estimate

$$\begin{aligned} &\mathsf{P}(S_V = N + u_1, T_V = E + u_2) \\ &\quad = |d| (2\pi V\Delta^{\frac{1}{2}})^{-1} e^{-(1/2\Delta V)(b_{11}u_2^2 - 2b_{12}u_1u_2 + b_{22}u_1^2)} \\ &\qquad + (m_0 + m_1u_1 + m_2u_2)V^{-2} + O[V^{-3}(V^{\frac{1}{2}} + u^3)], \end{aligned} \tag{36}$$

where m_0, m_1, m_2 are constants independent of V, u_1 and u_2.

If it is noted that

$$e^{-(1/2\Delta V)(b_{11}u_2^2 - 2b_{12}u_1u_2 + b_{22}u_1^2)} = 1 + O(V^{-1}u^2),$$

then equation (35) yields

$$\begin{aligned} &\mathsf{P}(S_V = N + u_1, T_V = E + u_2) \\ &\quad = (2\pi V\Delta^{\frac{1}{2}})^{-1}|d| + O[V^{-2}(1 + u^2)]. \end{aligned} \tag{37}$$

Similarly, if the more accurate estimate

$$\begin{aligned} &e^{-(1/2\Delta V)(b_{11}u_2^2 - 2b_{12}u_1u_2 + b_{22}u_1^2)} \\ &\quad = 1 - (1/2\Delta V)(b_{11}u_2^2 - 2b_{12}u_1u_2 + b_{22}u_1^2) + O(V^{-2}u^4) \end{aligned}$$

is used, then equation (36) yields

$$\begin{aligned} &\mathsf{P}(S_V = N + u_1, T_V = E + u_2) \\ &\quad = (2\pi V\Delta^{\frac{1}{2}})^{-1}|d| + V^{-2}\{m_0 + m_1u_1 + m_2u_2 \\ &\qquad - (4\pi\Delta^{\frac{3}{2}})^{-1}|d| (b_{11}u_2^2 - 2b_{12}u_1u_2 + b_{22}u_1^2)\} \\ &\qquad + O[V^{-3}(V^{\frac{1}{2}} + u^4)]. \end{aligned} \tag{38}$$

Formulas (37) and (38) will be used in all of the following sections. Let us state again that for these formulas to be valid we must choose values of u_1 and u_2 such that the point $(N + u_1, E + u_2)$ belongs to the lattice $Va_0 + k\alpha + l\beta$, $Vb_0 + k\gamma + l\delta$. In particular, this condition is always satisfied if $\Omega(N + u_1, E + u_2) > 0$, i.e., if the system composed of $N + u_1$ particles can have the total energy $E + u_2$.

§6. Mean values of sum functions

For the purposes of the present section we can limit ourselves to the crude estimate given by the very simple formula (37). Only later will we need the more accurate estimate given by formula (38).

First note that the theorem of §3 [formula (22)], in conjunction with (37), yields

$$\Omega(N + u_1, E + u_2)$$
$$= C(N + u_1)\Phi(\alpha, \beta)e^{\alpha N+\beta E}e^{\alpha u_1+\beta u_2}\{(2\pi V\Delta^{\frac{1}{2}})^{-1}|d| + O[V^{-2}(1 + u^2)]\}.$$

In particular, for $u_1 = u_2 = 0$, it follows from this that

$$\Omega(N, E) = C(N)\Phi(\alpha, \beta)e^{\alpha N+\beta E}\{(2\pi V\Delta^{\frac{1}{2}})^{-1}|d| + O(V^{-2})\},$$

and, therefore, that

$$\Omega(N + u_1, E + u_2)/\Omega(N, E) \tag{39}$$
$$= \{C(N + u_1)e^{\alpha u_1+\beta u_2}/C(N)\}\{1 + O[V^{-1}(1 + u^2)]\}.$$

Hence, having an asymptotic estimate of a ratio of the form

$$\Omega(N + u_1, E + u_2)/\Omega(N, E),$$

we can easily find very simple approximate expressions for the mean values of the occupation numbers on the basis of the formulas derived in §2.

In the case of complete statistics for which $C(p) = p!$, formula (6) of §2, together with formula (39), yields

$$<a_r> = e^{-(\alpha+\beta\varepsilon_r)}\{1 + O[V^{-1}(1 + \varepsilon_r^2)]\}. \tag{40}$$

In the case of symmetric statistics we have $C(p) = 1$, and formula (10) of §2, together with formula (39), yields

$$<a_r> = \sum_{k=1}^{\infty} e^{-k(\alpha+\beta\varepsilon_r)}\{1 + O(V^{-1}k^2\varepsilon_r^2)\} \tag{41}$$
$$= (e^{\alpha+\beta\varepsilon_r} - 1)^{-1}\{1 + O(V^{-1}\varepsilon_r^2 \sum_{k=1}^{\infty} k^2e^{-k(\alpha+\beta\varepsilon_r)})\}.$$

Also, in the case of antisymmetric statistics $C(p) = 1$, and formula (13) of §2, together with (39), yields

$$<a_r> = (e^{\alpha+\beta\varepsilon_r} + 1)^{-1}\{1 + O(V^{-1}\varepsilon_r^2 \sum_{k=1}^{\infty} k^2e^{-k(\alpha+\beta\varepsilon_r)})\}. \tag{42}$$

Evidently, (40), (41) and (42) can be combined into one very simple formula for any fixed value of r:

$$<a_r> = (e^{\alpha+\beta\varepsilon_r} - \sigma)^{-1} + O(V^{-1}), \tag{43}$$

where σ is the index of symmetry of the system.

In all these derivations, we relied on formula (39), which in turn was based on formula (37) of §5. Therefore, we must still convince ourselves that if (N, E) is a possible combination of the number of particles and of the total energy of the system, i.e., if the point (N, E) belongs to the lattice $Va_0 + k\alpha + l\beta$, $Vb_0 + k\gamma + l\delta$, then the point $(N - k, E - k\varepsilon_r)$ belongs to the same lattice for any integer $k > 0$. But this is obvious since $(1, \varepsilon_r)$ is a possible combination of the number of particles and the total energy of the system. Thus, if the points (N, E) and $(1, \varepsilon_r)$ both belong to the aforementioned lattice, so also does the point $(N - k, E - k\varepsilon_r)$ for any integer k.

Now let $\mathfrak{A}$ be a sum function related to the system, i.e., let

$$\mathfrak{A} = \sum_{i=1}^{N} \mathfrak{A}_i ,$$

where the quantity $\mathfrak{A}_i$ depends only on the Hamiltonian variables of the ith particle. If we assume that the mathematical expectation $\mathsf{E}_r\mathfrak{A}_i = \lambda_r$ of the quantity $\mathfrak{A}_i$, in the state corresponding to the energy level ε_r, is the same for all i and, consequently, depends only on r, then for the mean value of the quantity $\mathsf{E}\mathfrak{A}$ we have, according to III, §5

$$<\mathsf{E}\mathfrak{A}> = \sum_{r=1}^{\infty} \lambda_r <a_r>.$$

Therefore, (40), (41) and (42) yield

$$<\mathsf{E}\mathfrak{A}> = \sum_{r=1}^{\infty} \lambda_r (e^{\alpha+\beta\varepsilon_r} - \sigma)^{-1} + O(V^{-1} \sum_{r=1}^{\infty} \varepsilon_r^{\,2} \mid \lambda_r \mid \sum_{k=1}^{\infty} k^2 e^{-k(\alpha+\beta\varepsilon_r)}).$$

In view of the remark we made at the beginning of §3, every absolutely convergent series of the form

$$\sum_{r=1}^{\infty} f(\varepsilon_r)$$

is a quantity proportional to V. Therefore, we obtain

$$<\mathsf{E}\mathfrak{A}> = \sum_{r=1}^{\infty} \lambda_r (e^{\alpha+\beta\varepsilon_r} - \sigma)^{-1} + O(1),$$

if all the series on the right side of the next to the last equality converge absolutely. [This requires that the quantity $\mid \lambda_r \mid$ should not increase too rapidly — a condition which is always met in physical problems. We also assume that the double series is $O(V)$ (see below).]

In the majority of cases λ_r has the same value for all states with the same energy level ε_r, i.e., λ_r is a single-valued function $\varphi(\varepsilon_r)$ of ε_r. Hence, we can rewrite the last formula in terms of the numbers g_k which we introduced at the beginning of §3:

$$<\mathsf{E}\mathfrak{A}> = V \sum_{k=1}^{\infty} g_k \, \varphi(k) (e^{\alpha+\beta k} - \sigma)^{-1} + O(1).$$

This relation shows that $<\mathsf{E}\mathfrak{A}>$ is asymptotically proportional to V. (This is clear directly since V is proportional to the number of particles

N.) Moreover, this relation shows the remarkable accuracy that use of the limit theorems of the theory of probability gives even for a very crude estimate of the remainder terms. This accuracy can be increased substantially, as we shall soon see, if a more precise estimate of the remainder terms in the limit theorems is used in the computations.

§7. Correlation between occupation numbers

Our next problem is to estimate the dispersion of sum functions. To do this we must first estimate the correlation between occupation numbers with different subscripts. This is necessary since III, §6, (24) shows that the value of the dispersion depends in an essential way on the order of smallness of the difference $\langle a_r a_s \rangle - \langle a_r \rangle \langle a_s \rangle$. It can be foreseen, as in the photon case, that this correlation is very weak since the number of particles is very large and their mutual correlation depends only on the relations

$$\sum_{r=1}^{\infty} a_r = N, \qquad \sum_{r=1}^{\infty} a_r \varepsilon_r = E.$$

As before, we see that the main terms cancel when $\langle a_r \rangle \langle a_s \rangle$ is subtracted from $\langle a_r a_s \rangle$. Consequently, the degree of accuracy to which the formulas of §6 were derived is inadequate, and we must carry out all the computations anew, basing them this time on the more accurate formula (38) rather than on (37). First, (22) and (38) give for all three statistics

$$\begin{aligned}
&\Omega(N + u_1, E + u_2)\\
&\quad = C(N + u_1)\Phi(\alpha, \beta)e^{\alpha N+\beta E}e^{\alpha u_1+\beta u_2}\{(|d|/2\pi V\Delta^{\frac{1}{2}})\\
&\quad + V^{-2}[(m_0 + m_1u_1 + m_2u_2) - (|d|/4\pi\Delta^{\frac{3}{2}})(b_{22}u_1^2 - 2b_{12}u_1u_2 + b_{11}u_2^2)]\\
&\quad + O[V^{-3}(V^{\frac{1}{2}} + u^4)]\};
\end{aligned}$$

and, in particular, for $u_1 = u_2 = 0$,

$$\Omega(N, E) = C(N)\Phi(\alpha, \beta)e^{\alpha N+\beta E}\{(|d|/2\pi V\Delta^{\frac{1}{2}}) + m_0V^{-2} + O(V^{-\frac{5}{2}})\}.$$

Whence

$$\begin{aligned}
&\Omega(N + u_1, E + u_2)/\Omega(N, E)\\
(44)\quad &= [C(N + u_1)/C(N)]e^{\alpha u_1+\beta u_2}\{1 + V^{-1}[l_1u_1 + l_2u_2\\
&\qquad -(1/2\Delta)\ (b_{22}u_1^2 - 2b_{12}u_1u_2 + b_{11}u_2^2)] + O[V^{-2}(V^{\frac{1}{2}} + u^4)]\},
\end{aligned}$$

where l_1 and l_2 are independent of V, u_1 and u_2.

This more accurate formula must replace formula (39) of the preceding section in subsequent computations. For convenience in writing, we put

$$T_r = T_r(\alpha, \beta) = (e^{\alpha+\beta\varepsilon_r} - \sigma)^{-1}$$

for all three statistics.

Now we find the more precise expression for the mean values of the occupation numbers.

In the case of complete statistics formula (6) of §2 combined with (44) yields

$$\langle a_r \rangle = e^{-(\alpha+\beta\varepsilon_r)}\{1 - V^{-1}[l_1 + l_2\varepsilon_r + (1/2\Delta)\,(b_{22} - 2b_{12}\varepsilon_r + b_{11}\varepsilon_r^{\,2})] + O[V^{-2}(V^{\frac{1}{2}} + \varepsilon_r^{\,4})]\}.$$

Noting that in the case of complete statistics

$$T_r = e^{-(\alpha+\beta\varepsilon_r)},$$

and therefore that

$$\partial T_r/\partial\alpha = -\,e^{-(\alpha+\beta\varepsilon_r)}, \qquad \partial T_r/\partial\beta = -\,\varepsilon_r e^{-(\alpha+\beta\varepsilon_r)},$$

$$\partial^2 T_r/\partial\alpha^2 = e^{-(\alpha+\beta\varepsilon_r)}, \qquad \partial^2 T_r/\partial\alpha\partial\beta = \varepsilon_r e^{-(\alpha+\beta\varepsilon_r)},$$

$$\partial^2 T_r/\partial\beta^2 = \varepsilon_r^{\,2} e^{-(\alpha+\beta\varepsilon_r)},$$

we obtain

$$\langle a_r \rangle = T_r - V^{-1}[-l_1\partial T_r/\partial\alpha - l_2\partial T_r/\partial\beta + (1/2\Delta)(b_{22}\partial^2 T_r/\partial\alpha^2 - 2b_{12}\partial^2 T_r/\partial\alpha\partial\beta + b_{11}\partial^2 T_r/\partial\beta^2)] + O[V^{-2}(T_r V^{\frac{1}{2}} + \partial^4 T_r/\partial\beta^4)]. \tag{45}$$

In the case of symmetric statistics, formula (10) of §2 combined with (44) yields

$$\langle a_r \rangle = \textstyle\sum_{k=1}^{\infty} e^{-k(\alpha+\beta\varepsilon_r)}\{1 - V^{-1}[k(l_1 + l_2\varepsilon_r) + (k^2/2\Delta)(b_{22} - 2b_{12}\varepsilon_r + b_{11}\varepsilon_r^{\,2})] + O[V^{-2}(V^{\frac{1}{2}} + k^4\varepsilon_r^{\,4})]\}. \tag{46}$$

This time we have

$$T_r = (e^{\alpha+\beta\varepsilon_r} - 1)^{-1} = \textstyle\sum_{k=1}^{\infty} e^{-k(\alpha+\beta\varepsilon_r)},$$

from which it follows that

$$\partial T_r/\partial\alpha = -\textstyle\sum_{k=1}^{\infty} k e^{-k(\alpha+\beta\varepsilon_r)},$$

$$\partial T_r/\partial\beta = -\textstyle\sum_{k=1}^{\infty} k\varepsilon_r e^{-k(\alpha+\beta\varepsilon_r)},$$

$$\partial^2 T_r/\partial\alpha^2 = \textstyle\sum_{k=1}^{\infty} k^2 e^{-k(\alpha+\beta\varepsilon_r)},$$

$$\partial^2 T_r/\partial\alpha\partial\beta = \textstyle\sum_{k=1}^{\infty} k^2\varepsilon_r e^{-k(\alpha+\beta\varepsilon_r)},$$

$$\partial^2 T_r/\partial\beta^2 = \textstyle\sum_{k=1}^{\infty} k^2\varepsilon_r^{\,2} e^{-k(\alpha+\beta\varepsilon_r)}.$$

Therefore, we find

$$\langle a_r \rangle = T_r - V^{-1}[-l_1\partial T_r/\partial\alpha - l_2\partial T_r/\partial\beta + (1/2\Delta)(b_{22}\partial^2 T_r/\partial\alpha^2 - 2b_{12}\partial^2 T_r/\partial\alpha\partial\beta + b_{11}\partial^2 T_r/\partial\beta^2)] + O[V^{-2}(T_r V^{\frac{1}{2}} + \partial^4 T_r/\partial\beta^4)],$$

i.e., we find exactly the same formula (45) as in the case of complete statistics.

Finally, in the case of antisymmetric statistics, formula (13) of §2 combined with (44) yields a formula which differs from (46) only in that the kth term of the sum is multiplied by $(-1)^{k-1}$ $(k = 1, 2, \cdots)$. In this case since

$$T_r = (e^{\alpha+\beta\varepsilon_r} + 1)^{-1} = \sum_{k=1}^{\infty} (-1)^{k-1} e^{-k(\alpha+\beta\varepsilon_r)},$$

it is easy to see that (45) is also valid for the case of antisymmetric statistics. Hence, formula (45), which is the necessary improved asymptotic expression for $\langle a_r \rangle$, is valid for all three statistics. (It is understood, of course, that T_r has different values in each of the three cases.)

Now, let us estimate the quantity $\langle a_r a_s \rangle$ $(r \neq s)$. In the case of complete statistics, we must start from formula (7) of §2 which, combined with (44), yields

$$\begin{aligned}
\langle a_r a_s \rangle &= e^{-(\alpha+\beta\varepsilon_r)} e^{-(\alpha+\beta\varepsilon_s)} \{1 - V^{-1}[(l_1 + l_2\varepsilon_r) \\
&\quad + (l_1 + l_2\varepsilon_s) + (1/2\Delta)(4b_{22} - 4b_{12}[\varepsilon_r + \varepsilon_s] \\
&\quad + b_{11}[\varepsilon_r + \varepsilon_s]^2)] + O[V^{-2}(V^{\frac{1}{2}} + \varepsilon_r^{\,4} + \varepsilon_s^{\,4})]\} \\
&= T_r T_s - V^{-1}[-l_1 \partial(T_r T_s)/\partial\alpha - l_2 \partial(T_r T_s)/\partial\beta \\
&\quad + (1/2\Delta)(b_{22}\partial^2(T_r T_s)/\partial\alpha^2 - 2b_{12}\partial^2(T_r T_s)/\partial\alpha\partial\beta \\
&\quad + b_{11}\partial^2(T_r T_s)/\partial\beta^2)] \\
&\quad + O[V^{-2}(T_r T_s V^{\frac{1}{2}} + T_r\partial^4 T_s/\partial\beta^4 + T_s\partial^4 T_r/\partial\beta^4)].
\end{aligned} \tag{47}$$

Since the method of obtaining formulas of this kind has been illustrated in several examples, we can leave it to the reader to show independently that formula (47), which we derived for the case of complete statistics, is also valid for the other two statistical schemes.

Now we shall estimate the difference $\langle a_r a_s \rangle - \langle a_r \rangle \langle a_s \rangle$. First, note that formula (45) yields

$$\begin{aligned}
\langle a_r \rangle \langle a_s \rangle &= T_r T_s - V^{-1}\{-l_1 \partial(T_r T_s)/\partial\alpha - l_2 \partial(T_r T_s)/\partial\beta \\
&\quad + (1/2\Delta)[b_{22}(T_r\partial^2 T_s/\partial\alpha^2 + T_s\partial^2 T_r/\partial\alpha^2) \\
&\quad - 2b_{12}(T_r\partial^2 T_s/\partial\alpha\partial\beta + T_s\partial^2 T_r/\partial\alpha\partial\beta) \\
&\quad + b_{11}(T_r\partial^2 T_s/\partial\beta^2 + T_s\partial^2 T_r/\partial\beta^2)]\} \\
&\quad + O[V^{-2}(T_r T_s V^{\frac{1}{2}} + T_r\partial^4 T_s/\partial\beta^4 + T_s\partial^4 T_r/\partial\beta^4)].
\end{aligned} \tag{48}$$

Furthermore, subtracting (48) term by term from (47), we find

$$
\begin{aligned}
<a_r a_s> - <a_r><a_s> = {} & -(1/\Delta V)\{b_{22}(\partial T_r/\partial\alpha)(\partial T_s/\partial\alpha) \\
& - b_{12}[(\partial T_r/\partial\alpha)(\partial T_s/\partial\beta) + (\partial T_s/\partial\alpha)(\partial T_r/\partial\beta)] \\
& + b_{11}(\partial T_r/\partial\beta)(\partial T_s/\partial\beta)\} \\
& + O[V^{-2}(T_r T_s V^{\frac{1}{2}} + T_r\partial^4 T_s/\partial\beta^4 + T_s\partial^4 T_r/\partial\beta^4)].
\end{aligned} \tag{49}
$$

This formula, like formulas (45), (47) and (48), holds for each of the three fundamental statistical schemes. It shows that the difference

$$<a_r a_s> - <a_r><a_s>,$$

which is a measure of the correlation between the numbers a_r and a_s $(r \neq s)$, is infinitely small for the assumed conditions and is asymptotically proportional to V^{-1}. Hence, the problem posed in this section has been solved completely.

§8. Dispersion of sum functions and the suitability of microcanonical averages

We shall now estimate the dispersion of sum functions which, as we know, is required to establish the suitability of their microcanonical averages. Besides the asymptotic estimates of the numbers $<a_r>$ and $<a_r a_s> - <a_r><a_s>$, which we have already found, we also need an estimate of the microcanonical dispersion of the number a_r, i.e., we must estimate the number $<a_r^2> - (<a_r>)^2$. We shall study this latter problem first.

We saw at the end of §2 [formula (20)] that for all three fundamental statistical schemes

$$<a_r^2> = <a_r> + (1+\sigma)\tau_r,$$

where

$$\tau_r = <a_r a_s> \qquad (r \neq s,\ \varepsilon_r = \varepsilon_s).$$

Therefore,

$$
\begin{aligned}
<a_r^2> - (<a_r>)^2 = {} & <a_r> + \sigma(<a_r>)^2 \\
& + (1+\sigma)[<a_r a_s> - <a_r><a_s>]_{r\neq s,\varepsilon_r=\varepsilon_s},
\end{aligned} \tag{50}
$$

since $<a_r> = <a_s>$ if $\varepsilon_r = \varepsilon_s$.

This formula solves completely the problem posed above since the asymptotic expressions for the numbers $<a_r>$ and $<a_r a_s> - <a_r><a_s>$ $(r \neq s)$ are given, respectively, by formulas (45) and (49) of the preceding section.

According to III, §6, (24), we have for the microcanonical dispersion of a sum function $\mathfrak{A}$ the expression

$$
\begin{aligned}
D(\mathfrak{A}) &= \textstyle\sum_{r=1}^{\infty} (\mu_r - \lambda_r^2) <a_r> \\
&\quad + \textstyle\sum_{r,s=1}^{\infty} \lambda_r \lambda_s [<a_r a_s> - <a_r><a_s>] \\
&= \textstyle\sum_{r=1}^{\infty} (\mu_r - \lambda_r^2) <a_r> \\
&\quad + \textstyle\sum_{r \neq s} \lambda_r \lambda_s [<a_r a_s> - <a_r><a_s>] \\
&\quad + \textstyle\sum_{r=1}^{\infty} \lambda_r^2 [<a_r^2> - (<a_r>)^2].
\end{aligned} \tag{51}
$$

Putting, for any r and s,

$$
c_{rs} = \begin{cases} <a_r a_s> - <a_r><a_s> & (r \neq s), \\ [<a_r a_s> - <a_r><a_s>]_{r \neq s, \varepsilon_r = \varepsilon_s} & (r = s), \end{cases}
$$

we can rewrite (50) in the form

$$
<a_r^2> - (<a_r>)^2 = <a_r> + \sigma(<a_r>)^2 + (1 + \sigma) c_{rr} .
$$

Inserting this expression in the right side of formula (51), we find

$$
\begin{aligned}
D(\mathfrak{A}) &= \textstyle\sum_{r=1}^{\infty} (\mu_r - \lambda_r^2) <a_r> + \sum_{r \neq s} \lambda_r \lambda_s c_{rs} \\
&\quad + \textstyle\sum_{r=1}^{\infty} \lambda_r^2 [<a_r> + \sigma(<a_r>)^2] + (1 + \sigma) \sum_{r=1}^{\infty} \lambda_r^2 c_{rr} \\
&= \textstyle\sum_{r \neq s} \lambda_r \lambda_s c_{rs} + \sum_{r=1}^{\infty} \lambda_r^2 c_{rr} \\
&\quad + \textstyle\sum_{r=1}^{\infty} \{\mu_r <a_r> + \sigma \lambda_r^2 [c_{rr} + (<a_r>)^2]\} \\
&= \textstyle\sum_{r,s=1}^{\infty} \lambda_r \lambda_s c_{rs} + \sum_{r=1}^{\infty} \{\mu_r <a_r> + \sigma \lambda_r^2 [c_{rr} + (<a_r>)^2]\}.
\end{aligned} \tag{52}
$$

Now turning to the asymptotic estimate of the quantity $D(\mathfrak{A})$, we note first that the right side of formula (49) evidently represents the quantity c_{rs} for any r and s (equal or unequal). For brevity, let us put $\sum_{r=1}^{\infty} \lambda_r T_r = Q = Q(\alpha, \beta)$. Using (49), we obtain

$$
\begin{aligned}
&\textstyle\sum_{r,s=1}^{\infty} \lambda_r \lambda_s c_{rs} \\
&= -(1/\Delta V)\{b_{22}(\partial Q/\partial \alpha)^2 - 2b_{12}(\partial Q/\partial \alpha)(\partial Q/\partial \beta) + b_{11}(\partial Q/\partial \beta)^2\} \\
&\qquad + O[V^{-2}(Q^2 V^{\frac{1}{2}} + Q \partial^4 Q/\partial \beta^4)].
\end{aligned}
$$

Let us recall (see the beginning of §3, for example) that every absolutely convergent series such as

$$
\textstyle\sum_{r=1}^{\infty} f(\varepsilon_r)
$$

is a quantity proportional to V. We have defined Q as precisely such a series since λ_r and T_r depend on ε_r. It is evident that any partial derivative of the function $Q(\alpha, \beta)$ has the form of such a series. Hence, the quantity Q in the right side of the last equality, and all its partial derivatives, are quantities proportional to V. This yields

$$\sum_{r,s=1}^{\infty} \lambda_r \lambda_s c_{rs} = C_1 V + O(V^{\frac{1}{2}}),$$

where C_1 is a constant.

Furthermore, by the same reasoning,

$$\sum_{r=1}^{\infty} \{\mu_r \langle a_r \rangle + \sigma \lambda_r^2 [c_{rr} + (\langle a_r \rangle)^2]\} = C_2 V,$$

where C_2 is another constant. Therefore, because of the constancy of the ratio V/N, formula (52) yields

$$D(\mathfrak{A}) = (C_1 + C_2)V + O(V^{\frac{1}{2}}) = CN + O(N^{\frac{1}{2}}).$$

Hence, the microcanonical dispersion of a sum function $\mathfrak{A}$ is asymptotically proportional to the number of particles N. We saw in III, §6 that compliance with the relation

$$D(\mathfrak{A}) = o(N^2),$$

is sufficient to establish the suitability of microcanonical averages. We see now that this relation is (more than) satisfied for a sum function. Moreover, we obtained a very practical method to estimate asymptotically the microcanonical dispersion. This is of substantial value to fluctuation theory in physics.

§9. Determination of the numbers g_k for structureless particles in the absence of external forces

By analogy with the procedure used for photons in §5 of the preceding chapter, we now determine for material particles the number Vg_k of possible energy levels included between k and $k + 1$ when the particles are enclosed in a vessel of volume V. We again assume that the state of a particle is determined by the Hamiltonian variables x, y, z, p_x, p_y, p_z and that no external forces act on the particle. Hence its total energy consists only of kinetic energy and the potential energy due to the vessel walls. We will assume that this latter energy vanishes within the vessel and is infinitely large outside. We again emphasize that the problem concerning us now has no relation to statistics. In particular, we can imagine that we are concerned with a single particle.

As was remarked in §5 of the preceding chapter, the energy ε of a (non-relativistic) material particle is related to its momentum p by the equation

$$\varepsilon = mc^2 + p^2/2m,$$

where m is the mass of the particle and c is the velocity of light in vacuo. Selecting the zero level of energy suitably, we can replace this relation by the simpler one

$$\varepsilon = p^2/2m,$$

which gives the expression

$$\mathcal{H} = -(h^2/8\pi^2 m)(\partial^2/\partial x^2 + \partial^2/\partial y^2 + \partial^2/\partial z^2)$$

for the energy operator $\mathcal{H}$.

Thus, the time-independent Schrödinger equation becomes

$$-(h^2/8\pi^2 m)(\partial^2 U/\partial x^2 + \partial^2 U/\partial y^2 + \partial^2 U/\partial z^2) = EU,$$

or

$$\partial^2 U/\partial x^2 + \partial^2 U/\partial y^2 + \partial^2 U/\partial z^2 = -(8\pi^2 mE/h^2)U. \tag{53}$$

(We take only the kinetic energy of the particle into account since the potential of the wall is zero within the vessel.) The solutions of this equation describe the stationary states of a particle corresponding to the energy level E.

We noted in §5 of the preceding chapter that a linear basis of the solutions of (53) can be constructed from functions of the form

$$U = C \sin(ax + \eta) \sin(by + \zeta) \sin(cz + \xi),$$

where in this case

$$a^2 + b^2 + c^2 = 8\pi^2 mE/h^2. \tag{54}$$

We again assume that the vessel enclosing this particle is the parallelopiped $0 \leq x \leq l_1$, $0 \leq y \leq l_2$, $0 \leq z \leq l_3$, $l_1 l_2 l_3 = V$. Since the quantity $|U(x, y, z)|^2$, which is proportional to the probability of finding the particle at the point (x, y, z) of the vessel, must be zero at the vessel walls, any of the conditions $x = 0$, $y = 0$, $z = 0$, $x = l_1$, $y = l_2$, $z = l_3$ imply $U = 0$. The first three conditions evidently lead to the requirement $\eta = \zeta = \xi = 0$, so that

$$U = C \sin ax \sin by \sin cz, \qquad a^2 + b^2 + c^2 = 8\pi^2 mE/h^2.$$

From the fact that $U = 0$ for $x = l_1$, it evidently follows that $a = n_1\pi/l_1$, where n_1 is an integer. Similarly, the last two conditions lead to the requirements $b = n_2\pi/l_2$, $c = n_3\pi/l_3$, where n_2 and n_3 are integers. Because of (54), the numbers n_1, n_2 and n_3 must be related in the following way:

$$n_1^2/l_1^2 + n_2^2/l_2^2 + n_3^2/l_3^2 = 8mE/h^2.$$

This means that the possible energy levels of the stationary states of a particle are precisely those numbers E which have the form

$$(h^2/8m)(n_1^2/l_1^2 + n_2^2/l_2^2 + n_3^2/l_3^2),$$

where n_1, n_2 and n_3 are integers which can be assumed non-negative. Each such triplet of numbers gives one of the linearly independent solutions of the

Schrödinger equation (53) corresponding to the given boundary conditions ($U = 0$ if at least one of the equalities $x = 0$, $y = 0$, $z = 0$, $x = l_1$, $y = l_2$, $z = l_3$ is satisfied).

Now let k be any integer. In view of the above statements, the number of linearly independent eigenfunctions of the energy operator corresponding to eigenvalues $E \leq k$ will equal the number of solutions of the inequality

$$(h^2/8m)(n_1^2/l_1^2 + n_2^2/l_2^2 + n_3^2/l_3^2) \leq k$$

for integral $n_1 \geq 0$, $n_2 \geq 0$, $n_3 \geq 0$. But this number is evidently asymptotically (for large k) equal to one eighth of the volume of the ellipsoid

$$(h^2/8m)(x^2/l_1^2 + y^2/l_2^2 + z^2/l_3^2) = k,$$

i.e., is equal to

$$R(k) = \tfrac{1}{8}\{\tfrac{4}{3}\pi l_1 l_2 l_3 (8mk/h^2)^{3/2}\} = \tfrac{8}{3}(2)^{1/2}\pi V(m^{3/2}/h^3)k^{3/2}.$$

The number of linearly independent stationary states with energy levels included between k and $k + dk$ is approximately

$$(55) \qquad R'(k)\, dk = 4\pi 2^{1/2} V(m^{3/2}/h^3)k^{1/2}\, dk,$$

where the interval dk must not be too small (but, understandably, it must be small compared with k) or the approximate formula obtained will not have real significance. On the other hand, in our other notation this number is equal to

$$V\sum_{r=k}^{k+dk} g_r,$$

so that

$$\sum_{r=k}^{k+dk} g_r \approx (4\pi 2^{1/2} m^{3/2}/h^3)k^{1/2}\, dk.$$

Consequently, the "average" value of g_r in the interval $k < r < k + dk$ equals

$$(4\pi 2^{1/2} m^{3/2}/h^3)k^{1/2}.$$

As in the photon case, we can put

$$g_k = (4\pi 2^{1/2} m^{3/2}/h^3)k^{1/2}$$

directly for the majority of computations, particularly for large values of k. Also, as in the photon case, this number must be doubled if the particles we consider have "spin".

Let us make the following remark. In classical physics, the Hamiltonian function of a particle within the vessel has the form

$$H = (1/2m)(p_x^2 + p_y^2 + p_z^2).$$

The volume $\mathbf{V}(k)$ of that part of the (six-dimensional) phase space of a particle in which $H < k$ is equal to the product of the volume V of the vessel and the volume of the sphere $H = k$. This latter volume is obviously equal to

$$\tfrac{4}{3}\pi(2mk)^{\frac{3}{2}}.$$

Hence,

$$\mathbf{V}(k) = \tfrac{4}{3}\pi V(2mk)^{\frac{3}{2}}.$$

Therefore, the volume of that part of the phase space where $k < H < k + 1$ equals

$$\mathbf{V}(k + 1) - \mathbf{V}(k) \approx \mathbf{V}'(k) = 2\pi V(2m)^{\frac{3}{2}}k^{\frac{1}{2}}.$$

Comparing this with formula (55), we find

$$Vg_k \approx [\mathbf{V}(k + 1) - \mathbf{V}(k)]/h^3.$$

This means that, on the average, the number of linearly independent stationary states with energy levels between k and $k + 1$ equals the volume of that part of the classical phase space of a particle for which

$$k < H < k + 1,$$

if the quantity h^3 is taken as the unit of volume. In other words, one stationary state is equivalent, in a well-known sense, to a "cell" of volume h^3 in the classical phase space.

Chapter VI

THERMODYNAMIC CONCLUSIONS

§1. The problems of statistical thermodynamics

The most important problem of statistical physics has always been the explanation of the laws of thermodynamics on the basis of a model of the atomic structure of matter. It is well-known that thermodynamics has reached a high level of theoretical development independently of these models. Its logical scheme has been reduced to a purely deductive (axiomatic) structure where a small number of fundamental principles play the role of axioms. From these principles all further laws are deduced by purely logical means. The fundamental principles (axioms) are understood as laws of nature found by experimental means. Therefore, the problem of basing the foundation of thermodynamics on a model of the structure of matter always necessitates deriving these "axioms" from the model. The problem can be considered as solved when the derivation of these principles has been completed.

How may this derivation be carried out on the basis of our statistical theory? In order to find the answer to this question, we must state a fundamental difficulty in principle that arises here. In "classical" (i.e., phenomenological, non-statistical) thermodynamics the variables characterizing a given system depend upon a very small number of fundamental variables such as energy, volume, temperature, pressure, entropy, etc. If the values of an independent subset of these variables are known, then the thermodynamic state of the system is considered to be uniquely determined. On the other hand, in the statistical theory, we can have a situation such that for given values of the thermodynamic variables there exists an enormous number of different states of the system which are compatible with these values. Each eigenfunction U of the energy operator belonging to the eigenvalue E of this operator, determines one of these possible states. Moreover, there is always an entire continuum of these eigenfunctions and even if we restrict ourselves to a linear basis of this continuum we will nevertheless have a very large, albeit finite, number $\Omega(N, E)$ of such states. In a typical situation in classical thermodynamics a quantity $\mathfrak{A}$ is uniquely determined by the values of the energy E and the volume V of the system. In the statistical theory this same quantity can have widely different values in different states which are compatible with the given E and V. However, if the statistical theory is to provide a logical basis for thermodynamic principles, it must certainly yield a unique value for such a quantity. Furthermore, the value must coincide with that given to the quantity in classical thermodynamics.

As we have mentioned repeatedly, there is only one solution to this difficulty: In the statistical theory, we must give some *mean value* $<\mathfrak{A}>$ for the value of the quantity $\mathfrak{A}$ which is to be compared with the corresponding quantity of classical thermodynamics. The average must be taken over all states U which are accessible to the system, and must correspond to the given energy level E (and also, if necessary, to given values of some other parameters). The principle of averaging can be chosen arbitrarily, but we are obliged to show the suitability of the averages given by the statistical theory, i.e., to show that in an overwhelming majority of the states the value of the variable $\mathfrak{A}$ is extremely close to $<\mathfrak{A}>$. When this does not happen, it is meaningless to compare the value given by the theory with either the value given by experiment or the value given by the classical theory. In Chapters IV and V, where the microcanonical principle of averaging was used, this proof of the suitability was actually demonstrated in detail for the most important cases.

Thus, the path we must follow is clear: For each of the variables of interest in classical thermodynamics, we must find a corresponding phase function in our statistical theory. Moreover, we must show that the microcanonical averages of these phase functions are subject to the same relationships as are the corresponding variables in classical thermodynamics.

§2. External parameters, external forces and their mean values

The Hamiltonian function of a physical system as well as of an individual particle can depend on a number of parameters which define the position or the state of external bodies. A change in the values of these parameters causes a change in the form of the energy operator and by the same token causes changes in the energy levels and the possible states of the system or particle. In the preceding development, the presence of this type of parameter in the expression for the potential energy of a system or particle was by no means excluded. In fact, when examining a system of particles enclosed within some vessel, we always considered the potential energy to depend on the volume V of this vessel. The volume V is the most important parameter of the type being discussed. Other similar examples of parameters are the variables defining the position or the state of the sources of external fields which act on the system. Until now, however, we have not paid any particular attention to the possible presence of this type of parameter. We were able to do this because the values of these parameters were assumed to be strictly constant. It was in this sense that we called our system isolated. Now, however, we must consider interactions between the given system and the bodies surrounding it for which this type of parameter undergoes changes. These changes are related to the work done by the particles of our system, and, therefore, are related to the change of energy both of the particles and of the entire system. Thus, a gas which is enclosed in a vessel of cylindrical form, upon expanding, generates work by the force of its particles which moves a piston and thereby changes the volume of the vessel.

We shall call the parameters just described the *external parameters* of the system. Thus, the Hamiltonian function of each particle depends, not only on the usual Hamiltonian variables of this particle, but also on a sequence of external parameters $\lambda_1, \lambda_2, \cdots, \lambda_s$. From a mathematical point of view, the external parameters are characterized by the fact that they have the same value for all particles. In fact, if desired, this latter property can be taken as the formal definition of an external parameter. [Let us also mention that the parameters α and β which were previously introduced, and are usually called the *inner* parameters of the system, have this same property (i.e., the same value for all particles): We know that for a given structure of the particles and for known values of the external parameters, the numbers α and β are uniquely determined if the number of particles N and the total energy E of the system are given.]

Let one of the particles of our system be in the rth stationary state (energy level ε_r), and let the elementary work dw performed by this particle produce a change $d\lambda_i$ in the external parameter λ_i. This work is accompanied by a change $d\varepsilon_r$ in the energy of the particle:

$$d\varepsilon_r = (\partial\varepsilon_r/\partial\lambda_i)\, d\lambda_i .$$

Since, due to the law of conservation of energy, $dw = -d\varepsilon_r$, it follows that

$$dw = -(\partial\varepsilon_r/\partial\lambda_i)\, d\lambda_i .$$

In mechanics the coefficient of $d\lambda_i$ in this expression for elementary work is called the *generalized force* with which the particle acts upon the external bodies "in the direction" of the parameter λ_i. Thus, for the above particle, this generalized force is equal to $-\partial\varepsilon_r/\partial\lambda_i$. (It should be understood that the possible energy levels ε_r of the particle depend on the values of the external parameters. In Chapters IV and V, for example, we saw that these levels depend in an essential manner on the volume occupied by the system.)

If the change $d\lambda_i$ of the parameter λ_i is due to the total work done by all the particles of the system, then the generalized force Λ_i, with which the whole system acts upon the external bodies "in the direction" of the parameter λ_i, is equal to the sum of the generalized forces of all the individual particles of the system. Let the system be in the basic state U which corresponds to a particular choice of the occupation numbers $a_r(U)$. We then have

$$\Lambda_i = -\sum_{r=1}^{\infty} a_r(U)(\partial\varepsilon_r/\partial\lambda_i).$$

This relation shows that the generalized external forces Λ_i are phase functions in the statistical theory (and are defined at least for the set of basic eigenfunctions). If the elementary work δW done by the whole system causes the changes $d\lambda_1, d\lambda_2, \cdots, d\lambda_s$ in the external parameters, then we have

$$\delta W = \sum_{i=1}^{s} \Lambda_i \, d\lambda_i = -\sum_{i=1}^{s} \{\sum_{r=1}^{\infty} a_r(U)(\partial \varepsilon_r/\partial \lambda_i)\} \, d\lambda_i .$$

Since

$$\sum_{r=1}^{\infty} a_r(U)\varepsilon_r = E$$

is evidently the total energy of the system,

$$\Lambda_i = -\sum_{r=1}^{\infty} a_r(U)(\partial \varepsilon_r/\partial \lambda_i) = -\partial E/\partial \lambda_i , \tag{1}$$

and we may write

$$\delta W = -\sum_{i=1}^{s} (\partial E/\partial \lambda_i) \, d\lambda_i .$$

It is necessary to bear in mind that $\partial E/\partial \lambda_i$ is a phase function defined by the relation (1), so that for given values of the total energy and the external parameters, $\partial E/\partial \lambda_i$ is still not determined, but depends on the state U in which the system is found.

The generalized external forces

$$\Lambda_i = -\partial E/\partial \lambda_i$$

are thus phase functions (and at the same time, evidently, *sum functions*). In virtue of the general methodological discussion in §1 we must therefore assume that the generalized forces considered in phenomenological thermodynamics have as their analogs in the statistical theory the *microcanonical averages* of these phase functions, i.e., the quantities

$$<\Lambda_i> = -<\partial E/\partial \lambda_i> = -\sum_{r=1}^{\infty} <a_r>(\partial \varepsilon_r/\partial \lambda_i). \tag{2}$$

Likewise, the classical elementary work performed by the system on the external bodies should be interpreted in our theory as the microcanonical average

$$<\delta W> = -\sum_{i=1}^{s} <\partial E/\partial \lambda_i> d\lambda_i = -\sum_{i=1}^{s} [\sum_{r=1}^{\infty} <a_r>(\partial \varepsilon_r/\partial \lambda_i) \, d\lambda_i].$$

We obtain approximate expressions for the variables $<\Lambda_i>$ if we substitute in (2) the approximate expressions

$$<a_r> \approx (e^{\alpha+\beta\varepsilon_r} - \sigma)^{-1},$$

of V, §6, where σ is the index of symmetry of the system. This gives

$$<\Lambda_i> \approx -\sum_{r=1}^{\infty} (\partial \varepsilon_r/\partial \lambda_i)(e^{\alpha+\beta\varepsilon_r} - \sigma)^{-1} \quad (i = 1, 2, \cdots, s). \tag{3}$$

We recall that in V, §4 we found for the function $\Phi(\alpha, \beta)$ the general expression

$$\Phi(\alpha, \beta) = \prod_{r=1}^{\infty} \{ \sum_{n=0}^{\infty} \gamma(n) e^{-n(\alpha+\beta\varepsilon_r)} \},$$

where

$$\gamma(n) = \begin{cases} 1/n! & \text{for complete statistics,} \\ 1 & \text{for symmetric statistics,} \\ 1 \ (n \leq 1) & \multirow{2}{*}{\text{for antisymmetric statistics.}} \\ 0 \ (n > 1) & \end{cases}$$

Thus, in the case of complete statistics,

$$\Phi(\alpha, \beta) = \prod_{r=1}^{\infty} \exp [e^{-(\alpha+\beta\varepsilon_r)}],$$

and

$$\ln \Phi(\alpha, \beta) = \sum_{r=1}^{\infty} e^{-(\alpha+\beta\varepsilon_r)}.$$

In the case of symmetric statistics

$$\Phi(\alpha, \beta) = \prod_{r=1}^{\infty} [1 - e^{-(\alpha+\beta\varepsilon_r)}]^{-1},$$

and

$$\ln \Phi(\alpha, \beta) = -\sum_{r=1}^{\infty} \ln (1 - e^{-(\alpha+\beta\varepsilon_r)}).$$

Finally, in the case of antisymmetric statistics

$$\Phi(\alpha, \beta) = \prod_{r=1}^{\infty} \{1 + e^{-(\alpha+\beta\varepsilon_r)}\},$$

and

$$\ln \Phi(\alpha, \beta) = \sum_{r=1}^{\infty} \ln (1 + e^{-(\alpha+\beta\varepsilon_r)}).$$

An elementary calculation based on these formulas shows that in all three cases the relation

$$\partial \ln \Phi(\alpha, \beta)/\partial\lambda_i = -\beta \sum_{r=1}^{\infty} (\partial\varepsilon_r/\partial\lambda_i)(e^{\alpha+\beta\varepsilon_r} - \sigma)^{-1}$$

is valid. Therefore, for all three cases formula (3) gives

$$<\Lambda_i> = \beta^{-1}(\partial \ln \Phi/\partial\lambda_i) \qquad (1 \leq i \leq s), \tag{4}$$

and, consequently,

$$<\delta W> = \sum_{i=1}^{s} <\Lambda_i> d\lambda_i = \beta^{-1} \sum_{i=1}^{s} (\partial \ln \Phi/\partial\lambda_i)\, d\lambda_i\, . \tag{5}$$

In Chapter V we saw that the number of particles N and the total energy E of the system are obtained by differentiating the function $\ln \Phi(\alpha,\beta)$ with respect to the parameters α and β. Now we see that the mean values of the generalized forces and of the elementary work done by the system are expressed simply by means of partial derivatives of this same function $\ln \Phi$ with respect to the appropriate external parameters.

Formulas (4) and (5), as we shall see, form the basis of all the thermodynamic conclusions to be drawn from our statistical theory.

§3. The definition of entropy and the deduction of the second law of thermodynamics

An expression for the function $\Phi(\alpha, \beta)$, which occupies such a prominent place in our theory, was established in the preceding section for the three basic statistical schemes. It depends, as we have seen, not only on α and β, but also on the external parameters $\lambda_1, \lambda_2, \cdots, \lambda_s$. The logarithmic derivatives of this function, with respect to its different arguments, give us expressions for the most important variables characterizing the state of the system, such as the number of particles N, the energy E and the external forces $\langle \Lambda_i \rangle$ $(i = 1, 2, \cdots, s)$. It is well-known that the so-called "characteristic function of Planck", sometimes called "the thermodynamic potential", possesses analogous properties in phenomenological thermodynamics. In the classical theory this function depends on the temperature of the system and on the external parameters. In our theory there are two "inner" parameters — α and β. However, in view of the assumed constancy of the number of particles N of the system, the relation

$$\partial \ln \Phi(\alpha, \beta)/\partial\alpha = -N$$

permits us to eliminate one of these parameters (e.g., α) from the expressions for the function $\ln \Phi(\alpha, \beta)$ itself, and from any of its partial derivatives. Thus, we may assume that the function $\ln \Phi(\alpha, \beta)$ and each of its partial derivatives depend only on the parameters $\beta, \lambda_1, \cdots, \lambda_s$. The analogy with the classical theory naturally leads to the assumption that the parameter β is uniquely related to the temperature of the system. All our knowledge about this parameter confirms the assumption. Thus, in all cases where the system consists of several components in thermal contact with each other, a naturally defined value of the parameter β exists which is common for all these components. The parameter β must be related, therefore, to a physical quantity which, in the case of thermal equilibrium among several systems, has the same value for all of the parts. In thermodynamics, the temperature is just such a quantity.

We recall now that the values of the parameters α and β were chosen (V, §4) such that the function

$$F(\alpha, \beta) = e^{\alpha N + \beta E}\Phi(\alpha, \beta)$$

has its minimum value at the point (α, β). The function $F(\alpha, \beta)$, like $\Phi(\alpha, \beta)$, depends on the external parameters $\lambda_1, \lambda_2, \cdots, \lambda_s$ as well as on α and β.

We have

$$\ln F(\alpha, \beta) = \alpha N + \beta E + \ln \Phi(\alpha, \beta),$$

and, consequently (in view of the constancy of N),

$$\begin{aligned} d \ln F(\alpha, \beta) &= N\, d\alpha + E\, d\beta + \beta\, dE + d \ln \Phi(\alpha, \beta) \\ &= N\, d\alpha + E\, d\beta + \beta\, dE + (\partial \ln \Phi / \partial \alpha)\, d\alpha \\ &\qquad + (\partial \ln \Phi / \partial \beta)\, d\beta + \textstyle\sum_{i=1}^{s} (\partial \ln \Phi / \partial \lambda_i)\, d\lambda_i\,. \end{aligned}$$

In view of the relations

$$\partial \ln \Phi / \partial \alpha = -N, \qquad \partial \ln \Phi / \partial \beta = -E$$

and formula (5), §2, we find

$$d \ln F(\alpha, \beta) = \beta[dE + <\delta W>]. \tag{6}$$

According to the laws of classical thermodynamics, the elementary increment dE of the energy of a system, caused by corresponding changes in the temperature and in the external parameters, is composed of the work performed on the system by the external bodies, which evidently is equal to $-\delta W$, and of the quantity of heat δQ which is received by the system. Thus,

$$dE = -<\delta W> + \delta Q.$$

In the above it should be understood that δQ, like $<\delta W>$, is not the total differential of any function of the parameters α, β, λ_i. However, the relation (6), which may be rewritten in the form

$$d \ln F(\alpha, \beta) = \beta\, \delta Q, \tag{7}$$

clearly shows that the product $\beta\, \delta Q$ is a total differential, i.e., that β is an integrating factor for δQ.

One of the most convenient and generally used formulations of the second law of thermodynamics can be stated as follows: There exists a function S of the temperature and the external parameters, and another function θ, depending only on the temperature of the system, such that

$$\theta\, \delta Q = dS. \tag{8}$$

In other words, δQ has an integrating factor depending only on the temperature of the system. The relation (7) coincides with the relation (8) if we set

$$\theta = \beta, \qquad S = \ln F(\alpha, \beta).$$

In thermodynamics the function θ is given by the expression $1/kT$, where T is the absolute temperature of the system, and k is the so-called Boltzmann constant. Therefore, in statistical physics, the physical meaning of the parameter β is always defined by the universal formula

$$\beta = 1/kT.$$

The function S is called the *entropy* of the system: it is one of the most important variables of thermodynamic theory. Relation (7) now assumes the form

$$\delta Q/T = k\, dS,$$

and turns out to be a direct expression of the second law of thermodynamics, which thus appears as a direct consequence of our statistical theory. In particular, the function $F(\alpha, \beta)$, which was introduced in Chapter V as an auxiliary mathematical tool, now acquires a most important physical meaning — the quantity $\ln F(\alpha, \beta)$ is the entropy of the system. In Chapter V we considered our special choice of values of the parameters α and β as a purely mathematical device designed to yield a simpler form for our asymptotic expressions. Here we see that this choice leads to important physical consequences, one of which we now introduce as an example.

Assume that we have two systems initially completely isolated from each other, and also from the surrounding world. We shall denote, respectively, by the indices 1 and 2 the variables relating to the first and second system, and leave without indices the variables relating to the system obtained from the union of the two given systems. It is assumed that sufficient time has elapsed for thermal equilibrium to be established in the combined system. Evidently, we will have

$$N = N_1 + N_2, \qquad E = E_1 + E_2, \qquad \Phi(\alpha, \beta) = \Phi_1(\alpha, \beta)\Phi_2(\alpha, \beta).$$

Hence, the entropy of the combined system is

$$\begin{aligned} S &= N\alpha + E\beta + \ln \Phi(\alpha, \beta) \\ &= [N_1\alpha + E_1\beta + \ln \Phi_1(\alpha, \beta)] + [N_2\alpha + E_2\beta + \ln \Phi_2(\alpha, \beta)]. \end{aligned} \tag{9}$$

But the function

$$\ln F_1(\alpha, \beta) = N_1\alpha + E_1\beta + \ln \Phi_1(\alpha, \beta)$$

has its minimum value for $\alpha = \alpha_1$, $\beta = \beta_1$, while the function $\ln F_2(\alpha, \beta)$ has its minimum value for $\alpha = \alpha_2$, $\beta = \beta_2$ (since the numbers α_1, β_1, α_2, β_2 are defined by this requirement). Therefore, relation (9) gives

$$S \geq \ln F_1(\alpha_1, \beta_1) + \ln F_2(\alpha_2, \beta_2) = S_1 + S_2,$$

i.e., *if two systems initially isolated from each other are brought into thermal contact, then after equilibrium is established the entropy of the combined system will never be less than the sum of the entropies initially possessed by the component systems.*

The equilibrium state is characterized by the fact that the parameters α and β take on the same values for both components. Further, in the relation $S \geq S_1 + S_2$ the equality sign will hold true if, and only if, $\alpha_1 = \alpha_2 = \alpha$, $\beta_1 = \beta_2 = \beta$.

We recall that the law of conservation of energy, which is the first law of thermodynamics, was established at the end of Chapter II in the form appropriate for quantum mechanics. This law, in contrast to the second, can be proved without statistical methods. As in classical mechanics, it is a simple consequence of the general laws of the evolution of physical systems in time (i.e., in the case of quantum physics a consequence of Schrödinger's equation).

On the basis of the two fundamental laws now established, thermodynamics can be constructed in a purely deductive fashion independently of any special model of the structure of matter. Thus, we have completely established a foundation for thermodynamics on the basis of our statistical theory.

Supplement I

THE STATISTICS OF HETEROGENEOUS SYSTEMS

Throughout the main part of this book we have restricted ourselves to homogeneous systems, i.e., to systems consisting of particles of identical structure. The purpose of this restriction was to simplify the formal apparatus so that the reader could concentrate on the conceptual bases of the method. However, without any changes in principle, our method can be used to describe the statistics of heterogeneous systems which consist of particles of several different types. The complications in the computational formulas which are caused by this transition are of a purely technical character. Fundamentally, the difference consists only in that in place of the two-dimensional limit theorems of the theory of probability analogous multi-dimensional theorems must be applied. The formulations, proofs and the conditions for their applicability correspond completely to those established in Chapter I for the one-dimensional and two-dimensional cases. As a rule, for systems consisting of particles of k different types, it is appropriate to use a $(k + 1)$-dimensional limit theorem. The remaining details of the methods do not differ from those described in the simple case of homogeneous systems. Of course, for the case of a heterogeneous system, particles of different structure can obey different statistics.

In the present supplementary section we shall discuss briefly systems consisting of material particles of two different types. The transition from two to three or more types is quite trivial and involves only a simple extension of the formulas for the case of two different types.

Let the system under study consist of N particles of two different types. As usual, we allow these particles (in particular, those of different type) to exchange energy freely. At the same time the interaction energy of the particles is assumed to be so insignificant that in all our calculations we can take the total energy of the system equal to the sum of the total energies of all its constituent particles.

Let the numbers of particles of the first and second types be equal to N_1 and N_2, respectively, $(N_1 + N_2 = N)$. We denote the possible energy levels for particles of the first type (in the usual order) by $\varepsilon_1, \varepsilon_2, \cdots$, and for particles of the second type, by $\eta_1, \eta_2, \cdots$. (Both spectra are assumed to be discrete. This is the same assumption we have made everywhere in the text.) For the sake of brevity we call the set of particles of the first (second) type, the first (second) *component* of the system. Let $U_1, U_2, \cdots$ be a complete orthogonal set of "admissible" basic eigenfunctions of the energy operator of the first component, and let $V_1, V_2, \cdots$ be an analogous set for the second component. In virtue of the general results of

Chapter III, the set of functions U_iV_j $(i, j = 1, 2, \cdots)$ is a complete orthogonal set of basic eigenfunctions of the energy operator for the whole system. In a completely natural fashion we extend to our heterogeneous system the concept of a *structure function*. This function, denoted by $\Omega(N_1, N_2, E)$, is the number of eigenfunctions of the form U_iV_j which belong to the eigenvalue E of the energy operator of the system. Since the energy of the system is equal to the sum of the energies of its components, we denote by $\Omega_1(N_1, E)$ and $\Omega_2(N_2, E)$ the structure functions (defined in the usual manner) of these components. Thus,

$$\Omega(N_1, N_2, E) = \sum_{x=0}^{\infty} \Omega_1(N_1, x)\Omega_2(N_2, E - x). \tag{1}$$

Assume now that it is known that the system is in some definite state U_iV_j. This means that the first component is in the state U_i, while the second is in the state V_j. Since the states U_i and V_j are fundamental, the values of all the "occupation numbers" are precisely determined, i.e., for arbitrary r and s the number a_r of particles of the first component in the state with energy level ε_r, and the number b_s of particles of the second component in the state with energy level η_s, have definite values. The numbers a_r and b_s always satisfy the relations

$$(K) \quad \sum_{r=1}^{\infty} a_r = N_1, \qquad \sum_{s=1}^{\infty} b_s = N_2, \qquad \sum_{r=1}^{\infty} a_r\varepsilon_r + \sum_{s=1}^{\infty} b_s\eta_s = E.$$

Conversely, if a choice of occupation numbers $a_r \geq 0$, $b_s \geq 0$ is given satisfying the relations (K), then to this choice, in general, there corresponds a definite number of states of the system of the form U_iV_j. By using the method described in III, §5, it is easy to show that this number is equal to

$$C_1(N_1)C_2(N_2)\left[\prod_{r=1}^{\infty} \gamma_1(a_r)\right]\left[\prod_{s=1}^{\infty} \gamma_2(b_s)\right]. \tag{2}$$

The notation introduced above is analogous to that used in the text: $C_1(N_1) = N_1!$ or $= 1$ depending upon whether the first component obeys complete statistics or either of the two other statistics; $C_2(N_2)$ has a similar meaning for the second component. The functions $\gamma_1(a_r)$ and $\gamma_2(b_s)$ are defined for the first and second component, respectively, in complete analogy with our previous function $\gamma(a)$. (Each of these functions also depends on the type of statistics obeyed by its component.)

In order to obtain the number $\Omega(N_1, N_2, E)$ of eigenfunctions of the form U_iV_j belonging to the eigenvalue E of the energy operator of the system, we must obviously sum the expression (2) over all possible choices of the numbers $a_r \geq 0$, $b_s \geq 0$, which satisfy the relations (K). Thus, we find

$$\Omega(N_1, N_2, E) = C_1(N_1)C_2(N_2)\sum_{(K)} \left\{\prod_{r=1}^{\infty} \gamma_1(a_r)\right\}\left\{\prod_{s=1}^{\infty} \gamma_2(b_s)\right\}. \tag{3}$$

This fundamental expression for the structure function is completely analogous to formula (14) of III, §5. Here, as there, it serves as the starting point for all further calculations. In V, §2 we expressed the mean values of the occupation numbers and their pairwise products in terms of ratios of structure functions. We will not carry out the analogous calculation here since, if necessary, the reader can independently establish the appropriate formulas.

However, we shall stop to examine, somewhat in detail, the second step of the investigation. This step is the reduction of the problem of finding an asymptotic estimate for the structure function to a limit problem of the theory of probability.

It is necessary to introduce two degenerate two-dimensional distribution laws u_k and v_l, where k and l are arbitrary integers. We also introduce three parameters α_1, α_2 and β, whose values are defined so that all the series to be considered will converge absolutely. The random vector (x, y), which is distributed according to the law u_k, has as its possible pairs of values only points of the form $x = n$, $y = nk$ $(n = 0, 1, \cdots)$, where

$$\mathsf{P}(x = n, y = nk)$$
$$= \gamma_1(n)e^{-n(\alpha_1+\beta k)}\left\{\sum_{m=0}^{\infty} \gamma_1(m)e^{-m(\alpha_1+\beta k)}\right\}^{-1} \qquad (n = 0, 1, \cdots).$$

Analogously, for the law v_l the possible pairs of values are the points of the form $x = n$, $y = nl$ $(n = 0, 1, \cdots)$, where

$$\mathsf{P}(x = n, y = nl)$$
$$= \gamma_2(n)e^{-n(\alpha_2+\beta l)}\left\{\sum_{m=0}^{\infty} \gamma_2(m)e^{-m(\alpha_2+\beta l)}\right\}^{-1} \qquad (n = 0, 1, \cdots).$$

For brevity, we denote, respectively, by g_{1k} and g_{2l} the degrees of degeneracy of the energy levels $\varepsilon_r = k$ and $\eta_s = l$ for particles of the first and second types, when the system occupies unit volume. We consider the sum of an infinite series of mutually independent random vectors (x_{1i}, y_{1i}) $(i = 1, 2, \cdots)$, among which there are g_{1k} vectors distributed according to the law u_k $(k = 1, 2, \cdots)$, and we set

$$\sum_{i=1}^{\infty} x_{1i} = X_1, \qquad \sum_{i=1}^{\infty} y_{1i} = Y_1.$$

It is easy to show that both series converge with probability 1, (see IV, §3). Similarly, let (x_{2i}, y_{2i}) $(i = 1, 2, \cdots)$ be a sequence of mutually independent random vectors, among which there are g_{2l} vectors distributed according to the law v_l $(l = 1, 2, \cdots)$, and set

$$\sum_{i=1}^{\infty} x_{2i} = X_2, \qquad \sum_{i=1}^{\infty} y_{2i} = Y_2.$$

Now we set $Y_1 + Y_2 = Y$ and we denote by P the distribution law of the three-dimensional vector (X_1, X_2, Y). Evidently this law depends only on the nature of the particles composing the system and on the parameters α_1, α_2 and β. In particular, it is independent of the numbers N_1, N_2, V

and E (if, as we shall assume here, the ratios of these numbers always maintain constant values).

Assume, finally, that we have the sum

$$(S_{1V}, S_{2V}, T_V) \tag{4}$$

of V mutually independent three-dimensional random vectors, each of which is distributed according to the law P, described above. Then, we may set

$$S_{1V} = \sum_{k=1}^{\infty} \sum_{i=1}^{Vg_{1k}} x_{1k_i}, \qquad S_{2V} = \sum_{l=1}^{\infty} \sum_{i=1}^{Vg_{2l}} x_{2l_i},$$
$$T_V = \sum_{k=1}^{\infty} \sum_{i=1}^{Vg_{1k}} y_{1k_i} + \sum_{l=1}^{\infty} \sum_{i=1}^{Vg_{2l}} y_{2l_i},$$

where the vector (x_{1k_i}, y_{1k_i}) obeys the law u_k, the vector (x_{2l_i}, y_{2l_i}) obeys the law v_l, and all these elementary vectors are mutually independent. We find, therefore, for the distribution law of the three-dimensional vector (4)

$$\begin{aligned} &\mathsf{P}(S_{1V} = p_1, S_{2V} = p_2, T_V = q) \\ &= \sum_{(K_{p_1 p_2 q})} \Big\{\Big[\prod_{k=1}^{\infty} \prod_{i=1}^{Vg_{1k}} \mathsf{P}(x_{1k_i} = a_{k_i}, y_{1k_i} = ka_{k_i})\Big] \\ &\qquad \cdot \Big[\prod_{l=1}^{\infty} \prod_{i=1}^{Vg_{2l}} \mathsf{P}(x_{2l_i} = b_{l_i}, y_{2l_i} = lb_{l_i})\Big]\Big\}, \end{aligned} \tag{5}$$

where the summation extends over all possible sets of numbers a_{k_i}, b_{l_i}, satisfying the relations

$$\begin{aligned} &\sum_{k=1}^{\infty} \sum_{i=1}^{Vg_{1k}} a_{k_i} = p_1, \qquad \sum_{l=1}^{\infty} \sum_{i=1}^{Vg_{2l}} b_{l_i} = p_2, \\ &\sum_{k=1}^{\infty} \sum_{i=1}^{Vg_{1k}} ka_{k_i} + \sum_{l=1}^{\infty} \sum_{i=1}^{Vg_{2l}} lb_{l_i} = q. \end{aligned} \tag{$K_{p_1p_2q}$}$$

According to the definition of the laws u_k and v_l, we have

$$\mathsf{P}(x_{1k_i} = a_{k_i}, y_{1k_i} = ka_{k_i}) = \Gamma_1(k)\gamma_1(a_{k_i})e^{-a_{k_i}(\alpha_1+\beta k)},$$
$$\mathsf{P}(x_{2l_i} = b_{l_i}, y_{2l_i} = lb_{l_i}) = \Gamma_2(l)\gamma_2(b_{l_i})e^{-b_{l_i}(\alpha_2+\beta l)},$$

where

$$\Gamma_1(k) = \Big\{\sum_{m=0} \gamma_1(m)e^{-m(\alpha_1+\beta k)}\Big\}^{-1},$$
$$\Gamma_2(l) = \Big\{\sum_{m=0} \gamma_2(m)e^{-m(\alpha_2+\beta l)}\Big\}^{-1}.$$

Substituting these expressions into the right side of relation (5), we find

$$\begin{aligned} &\mathsf{P}(S_{1V} = p_1, S_{2V} = p_2, T_V = q) \\ &= \Big\{\prod_{k=1}^{\infty} [\Gamma_1(k)]^{Vg_{1k}}\Big\}\Big\{\prod_{l=1}^{\infty} [\Gamma_2(l)]^{Vg_{2l}}\Big\} \sum_{(K_{p_1p_2q})} \Big[\prod_{k=1}^{\infty} \prod_{i=1}^{Vg_{1k}} \gamma_1(a_{k_i})\Big] \\ &\qquad \cdot \Big[\prod_{l=1}^{\infty} \prod_{i=1}^{Vg_{2l}} \gamma_2(b_{l_i})\Big] \exp\Big[-\alpha_1 \sum_{k=1}^{\infty} \sum_{i=1}^{Vg_{1k}} a_{k_i} - \alpha_2 \sum_{l=1}^{\infty} \sum_{i=1}^{Vg_{2l}} b_{l_i} \\ &\qquad\qquad - \beta\Big(\sum_{k=1}^{\infty} \sum_{i=1}^{Vg_{1k}} ka_{k_i} + \sum_{l=1}^{\infty} \sum_{i=1}^{Vg_{2l}} lb_{l_i}\Big)\Big] \\ &= e^{-\alpha_1 p_1 - \alpha_2 p_2 - \beta q}\Big\{\prod_{k=1}^{\infty} [\Gamma_1(k)]^{Vg_{1k}}\Big\}\Big\{\prod_{l=1}^{\infty} [\Gamma_2(l)]^{Vg_{2l}}\Big\} \\ &\qquad \cdot \sum_{(K_{p_1p_2q})} \Big[\prod_{k=1}^{\infty} \prod_{i=1}^{Vg_{1k}} \gamma_1(a_{k_i})\Big] \prod_{l=1}^{\infty} \prod_{i=1}^{Vg_{2l}} \gamma_2(b_{l_i}). \end{aligned}$$

The last sum on the right side of this relation differs only in notation from the sum

$$\sum_{(K)} [\prod_{r=1}^{\infty} \gamma_1(a_r)][\prod_{s=1}^{\infty} \gamma_2(b_s)],$$

which appears in the right side of formula (3). Thus,

$$\mathsf{P}(S_{1V} = p_1, S_{2V} = p_2, T_V = q)$$

$$= \{\prod_{k=1}^{\infty} [\Gamma_1(k)]^{Vg_{1k}}\}\{\prod_{l=1}^{\infty} [\Gamma_2(l)]^{Vg_{2l}}\} e^{-\alpha_1 p_1 - \alpha_2 p_2 - \beta q} \cdot \Omega(p_1, p_2, q)[C_1(p_1)C_2(p_2)]^{-1}. \tag{6}$$

Summing this equation over all integers p_1, p_2, q, we find

$$1 = \{\prod_{k=1}^{\infty} [\Gamma_1(k)]^{Vg_{1k}}\}\{\prod_{l=1}^{\infty} [\Gamma_2(l)]^{Vg_{2l}}\} \sum_{p_1, p_2, q} e^{-\alpha_1 p_1 - \alpha_2 p_2 - \beta q} \cdot \Omega(p_1, p_2, q)[C_1(p_1)C_2(p_2)]^{-1}.$$

The sum on the right side of this equation is a function of the parameters α_1, α_2 and β. We denote this sum by $\Phi(\alpha_1, \alpha_2, \beta)$, so that

$$\{\prod_{k=1}^{\infty} [\Gamma_1(k)]^{Vg_{1k}}\}\{\prod_{l=1}^{\infty} [\Gamma_2(l)]^{Vg_{2l}}\} = [\Phi(\alpha_1, \alpha_2, \beta)]^{-1}.$$

The relation (6) gives

$$\mathsf{P}(S_{1V} = p_1, S_{2V} = p_2, T_V = q)$$

$$= [\Phi(\alpha_1, \alpha_2, \beta)]^{-1} e^{-\alpha_1 p_1 - \alpha_2 p_2 - \beta q} \Omega(p_1, p_2, q)[C_1(p_1)C_2(p_2)]^{-1};$$

and, hence,

$$\Omega(p_1, p_2, q) = \Phi(\alpha_1, \alpha_2, \beta) C_1(p_1) C_2(p_2) e^{\alpha_1 p_1 + \alpha_2 p_2 + \beta q} \cdot \mathsf{P}(S_{1V} = p_1, S_{2V} = p_2, T_V = q).$$

This formula, in exact analogy with formula (22) of V, §3 reduces the problem of finding an asymptotic estimate for the structure function $\Omega(p_1, p_2, q)$ to a limit problem of the theory of probability, since the vector (S_{1V}, S_{2V}, T_V) is defined as the sum of V mutually independent identically distributed three-dimensional random vectors. It should be remarked that the heterogeneity of our system results only in an increase of the number of dimensions in the corresponding problem of the theory of probability. As previously, it is necessary to sum only random vectors which are identically distributed.

The last step — the application of the (three-dimensional) local limit theorem — we shall not consider since, in essence, the procedure does not differ from that described in the two-dimensional case. We mention only that the values of the parameters α_1, α_2 and β are determined from the relations

$$\partial \ln \Phi/\partial\alpha_1 = -N_1, \quad \partial \ln \Phi/\partial\alpha_2 = -N_2, \quad \partial \ln \Phi/\partial\beta = -E.$$

The existence and uniqueness of the solutions of these equations may be established by a procedure analogous to that used in Chapter V. Since $\ln \Phi(\alpha_1, \alpha_2, \beta)$ (and each of its partial derivatives) is proportional to V, and since the ratios N_1/V, N_2/V, E/V are assumed to be constant, the parameters α_1, α_2 and β must be constants.

In regard to the fundamental results of the theory, we remark that the mean values of the occupation numbers are given by the expressions

$$<a_r> = (e^{\alpha_1+\beta\varepsilon_r} - \sigma_1)^{-1} + O(V^{-1}),$$

and

$$<b_s> = (e^{\alpha_2+\beta\eta_s} - \sigma_2)^{-1} + O(V^{-1}),$$

where σ_1 and σ_2, respectively, denote the indices of symmetry of the first and second components. It should be understood that the energy E_1 is not a constant, but is a phase function of the system, and, in particular, a sum function. Its mean value can be written directly if the mean values of the occupation numbers are known. Thus, to a good approximation, we find

$$<E_1> \approx \sum_{r=1}^{\infty} \varepsilon_r (e^{\alpha_1+\beta\varepsilon_r} - \sigma_1)^{-1},$$

and, analogously, for the second component

$$<E_2> \approx \sum_{s=1}^{\infty} \eta_s (e^{\alpha_2+\beta\eta_s} - \sigma_2)^{-1}.$$

Supplement II

THE DISTRIBUTION OF A COMPONENT AND ITS ENERGY

§1

Let a system be composed of two components, and let us retain the terminology and the notation introduced in the preceding supplement. The states of this system, corresponding to an energy level E, constitute a linear manifold. A basis of this manifold is provided by a set of eigenfunctions of the form $U_i V_j$, where U_i is one of the fundamental eigenfunctions of the first component, belonging to some energy level E_1, and V_j is one of the fundamental eigenfunctions of the second component, belonging to some energy level E_2. The energy eigenvalues of the appropriate functions U_i and V_j must satisfy $E_1 + E_2 = E$. As we saw in the preceding supplement [formula (1)], these facts permit us to write down immediately the important relation

$$(1) \qquad \Omega(N_1, N_2, E) = \textstyle\sum_{x=0}^{\infty} \Omega_1(N_1, x)\Omega_2(N_2, E - x),$$

which will serve as the starting point for our further calculations.

First we find the number of those fundamental functions $U_i V_j$ of our system, belonging to the energy level E, in which the first index i has a definite value, i.e., in which the first component is found in a definite fundamental state U_i. If this state corresponds to the energy level $E_1 = E_1(U_i)$ of the first component, then the unknown quantity is obviously the number of functions V_j corresponding to the energy level $E - E_1$, i.e., the number

$$\Omega_2(N_2, E - E_1) = \Omega_2[N_2, E - E_1(U_i)].$$

Suppose now that $\mathfrak{A}$ is a physical quantity that has a definite value in each of the fundamental states $U_i V_j$ of this system (a phase function!). Moreover, let this quantity be completely determined by the state U_i of the first component, and hence be independent of the state V_j of the second component, so that

$$\mathfrak{A} = f(U_i).$$

Then this quantity retains the same value in all of those $\Omega_2[N_2, E - E_1(U_i)]$ fundamental states $U_i V_j$ of the system in which the first factor is equal to U_i (i.e., in which the first component is in the state U_i). Thus the microcanonical average of such a quantity can be written in the form

$$\begin{aligned} <\mathfrak{A}> &= <f(U_i)> \\ &= [\Omega(N_1, N_2, E)]^{-1}\textstyle\sum_{i=1}^{\infty} f(U_i)\Omega_2[N_2, E - E_1(U_i)]. \end{aligned} \tag{2}$$

This relation, being almost self-evident, nonetheless has a profound meaning and, as we shall see later, important consequences. First, it shows that for quantities which depend only on the first component, the microcanonical average (which is always an average over all the fundamental functions U_iV_j of our total system belonging to the energy level E) can be replaced *by a certain average over the fundamental eigenfunctions* U_i *of the first component.* We note immediately, however, that this "reduced" average has features which sharply distinguish it from the microcanonical average: In the microcanonical average only states with the same fixed energy participate, while the definition (2) includes all fundamental eigenfunctions U_i of the first component, regardless of the energy levels to which they might belong. Further, in the microcanonical average all participating states have equal weight while in (2) the fundamental state U_i has the weight

$$\Omega_2[N_2, E - E(U_i)]/\Omega(N_1, N_2, E),$$

which depends on the value $E_1(U_i)$ of the energy of the first component in the state U_i, and may therefore be different for different fundamental states U_i.

We consider now the important and frequently encountered special case in which the quantity $\mathfrak{A}$ depends only on the energy E_1 of the first component, i.e., for all states U_i, corresponding to the same energy level E_1, it has the same value

$$\mathfrak{A} = \varphi(E_1) = \varphi[E_1(U_i)].$$

Then in formula (2), $f(U_i) = \varphi(x)$ in all terms for which $E_1(U_i) = x$. The number of such terms is the number of fundamental eigenfunctions U_i of the first component, corresponding to the energy level x, i.e., $\Omega_1(N_1, x)$. Thus, formula (2) assumes the form

$$\begin{aligned} <\mathfrak{A}> &= <\varphi(E_1)> \\ &= [\Omega(N_1, N_2, E)]^{-1}\textstyle\sum_{x=0}^{\infty} \varphi(x)\Omega_1(N_1, x)\Omega_2(N_2, E - x). \end{aligned} \tag{3}$$

In particular, the microcanonical average of the energy of the first component is

$$<E_1> = [\Omega(N_1, N_2, E)]^{-1}\textstyle\sum_{x=0}^{\infty} x\Omega_1(N_1, x)\Omega_2(N_2, E - x). \tag{4}$$

Formulas (2) — (4) are exact. However, for actual calculations they are completely useless because of the complexity of the expressions for the functions Ω_1, Ω_2 and Ω. Therefore, for practical computations, we must replace them by simpler approximate formulas, which can easily be obtained with the help of the method developed in this book. We now study several important applications of this method. In most cases, we limit ourselves to the derivation of asymptotic expressions, and do not pause to make detailed estimates of the omitted terms.

§2

To obtain the necessary asymptotic expressions, we first turn to V, §6, (39). If we replace Ω by Ω_2 and N by N_2 in this formula, and assume $u_1 = 0$, $u_2 = -x$, then we find

$$\Omega_2(N_2, E - x)/\Omega_2(N_2, E) = e^{-\beta x}\{1 + O[V^{-1}(1 + x^2)]\}.$$

Applying this estimate to the weight function

$$\Omega_2[N_2, E - E_1(U_i)]/\Omega(N_1, N_2, E)$$
$$= \Omega_2[N_2, E - E_1(U_i)][\textstyle\sum_{x=0}^{\infty} \Omega_1(N_1, x)\Omega_2(N_2, E - x)]^{-1}$$

of formula (2), we easily find for it the very simple asymptotic expresssion

$$e^{-\beta E_1(U_i)}[\textstyle\sum_{x=0}^{\infty} \Omega_1(N_1, x)e^{-\beta x}]^{-1} + O(V^{-1}).$$

Thus, formula (2) can be replaced by the approximate formula

$$(5) \quad <\mathfrak{A}> = <f(U_i)> \approx \textstyle\sum_{i=1}^{\infty} f(U_i)e^{-\beta E_1(U_i)}[\sum_{x=0}^{\infty} \Omega_1(N_1, x)e^{-\beta x}]^{-1}.$$

Our derivation of this formula is very simple, but it contains an inaccuracy which makes formula (5) incorrect in the general case. Thus, in formula (39) of V, §6, upon which our calculation is based, E denotes the fixed total energy of a system whose structure function is $\Omega(N, E)$. But in our case E denotes the fixed value of the energy of our *total system*, while Ω_2 is the structure function of the *second component*. We shall, however, consider only systems in which the first component is negligibly small compared to the second. More precisely, we shall construct our asymptotic formulas under the assumption that the numbers N_2, E and V approach infinity, maintaining constant ratios, while the number N_1 remains constant. The most interesting application is the case $N_1 = 1$ for which the first component is an individual particle. In this case the energy E_2 of the second component is asymptotically equal to the energy E of the total system and, as can be seen without difficulty by a more detailed calculation, the error we mentioned above does not reduce the accuracy of equation (5).

For the case just described, we call the first component a *small component* of the system. Thus, *for quantities which are determined by the state of a small component* (in particular, of an individual particle), *the microcanoni-*

cal averages can be obtained approximately by averaging over all the fundamental functions of this component with the weight function

$$e^{-\beta E_1(U_i)}\left[\sum_{x=0}^{\infty} \Omega_1(N_1, x)e^{-\beta x}\right]^{-1}. \tag{6}$$

In statistical physics this result is called Boltzmann's law, and it is usual to consider the weight function (6) as the *probability* of finding the small component in the state U_i. There would be no objection to such a terminology if, in the subsequent development of the theory, the term "probability" always retained this meaning. However, experience has shown that on the basis of this terminology, as a rule, a good number of complications and imprecise formulations arise. Therefore, here, as in the rest of the book, we shall consciously avoid probabilistic terminology, especially since there is no pressing necessity for its use.

With the help of expression (6) for the weight function, formulas (3) and (4) can be rewritten as asymptotic expressions:

$$\begin{aligned} <\mathfrak{A}> &= <\varphi(E_1)> \\ &\approx \sum_{x=0}^{\infty} \varphi(x)\Omega_1(N_1, x)e^{-\beta x}\left[\sum_{x=0}^{\infty} \Omega_1(N_1, x)e^{-\beta x}\right]^{-1} \end{aligned} \tag{7}$$

and

$$<E> \approx \sum_{x=0}^{\infty} x\Omega_1(N_1, x)e^{-\beta x}\left[\sum_{x=0}^{\infty} \Omega_1(N_1, x)e^{-\beta x}\right]^{-1}. \tag{8}$$

Thus, *the microcanonical average of any function of the energy of a small component* (in particular, of an individual particle) *can be obtained approximately by averaging the function over all possible values x of this energy with the weight function*

$$\Omega_1(N_1, x)e^{-\beta x}\left[\sum_{x=0}^{\infty} \Omega_1(N_1, x)e^{-\beta x}\right]^{-1}.$$

Supplement III

THE PRINCIPLE OF CANONICAL AVERAGING

In this book we have always assumed that the system under study was energetically isolated, i.e., not exchanging energy with surrounding bodies. The whole method of microcanonical averaging is constructed on this premise, because the only states which participate in the formation of microcanonical averages are states in which the energy of the system has a strictly fixed value. However, this requirement of complete energetic isolation can be achieved only approximately under real conditions. In fact, in many cases of practical importance, the system is in more or less intensive energetic (for example, thermal) contact with surrounding bodies. As a consequence, its energy does not remain constant and, hence, the principle of microcanonical averaging is deprived of its theoretical foundation.

The results of Supplement II permit us to use our statistical theory in a rational way to construct the statistics of such non-isolated systems. We consider the extreme case in which the system can freely exchange energy with its almost infinitely large environment (whose energy is many times larger than that of our system). Physicists call such an environment a "heat bath", and say that the system is "immersed in a heat bath". Theoretical considerations lead to the conclusion that for a system immersed in a heat bath only the *temperature* of the heat bath is significant. Its other properties, including even its material composition (i.e., the nature of its component particles), are irrelevant. In particular, nothing prevents us from representing the heat bath in the form of an enormous number of physical systems which are exact replicas of the one being studied, and which freely exchange energy among themselves as well as with the given system.

If we adopt this point of view, then we can consider the combination of the given system S and the heat bath T as one isolated system $S + T$, with respect to which our system S, and each of the similar systems which comprise the heat bath T, play the role of individual particles. Since the system $S + T$ is isolated, we assume, of course, that all the preceding principles of microcanonical averaging are justified. Since the system S can obviously be considered as a small component (individual particle) of the isolated system $S + T$, a microcanonical average over the states of the system $S + T$ is equivalent for system S to an average according to formula (2) of Supplement II. In this formula Ω_2 and Ω denote the structure functions of systems T and $S + T$, respectively; $N_1 = 1$ and N_2 is equal to the number of systems (identical to S) which constitute the heat bath. Also, E is the (constant) total energy of the combined system $S + T$,

$E_1(U_i)$ is the energy of the given system in the fundamental state U_i, and the summation extends over all such states. This conclusion is exact. However, as we have seen, formula (2) can to a good approximation be replaced by the incomparably simpler and more convenient formula (5), where the summation extends over the same range and where Ω_1 is the structure function of the system S. As for the parameter β, we know that it is universally related to the absolute temperature of the system by the equation

$$\beta = 1/kT,$$

where k is Boltzmann's constant (and where the symbol T for the absolute temperature, which we use here only in passing, is not to be confused with the symbol for the heat bath). Of course, the temperature of the system is the same as the temperature of the heat bath. It is clear that the principle of averaging for the system S, expressed by formula (5), is independent of the special nature of the heat bath and depends (through the parameter β) only on its temperature. This temperature is fixed, both for the system and for the heat bath, as a result of the energetic contact between them.

We see, therefore, that *if we accept microcanonical averaging as the primary basis for statistical computations for isolated systems, then this necessarily implies a certain definite principle of averaging for systems which are freely exchanging energy with a large environment (heat bath). This principle is expressed approximately by formula* (5) *of the preceding supplement*:

$$(1) \quad \langle \mathfrak{A} \rangle = \langle f(U_i) \rangle \approx \sum_{i=1}^{\infty} f(U_i) e^{-\beta E(U_i)} \left[\sum_{x=0}^{\infty} \Omega(N, x) e^{-\beta x} \right]^{-1},$$

where N is the number of particles, Ω is the structure function, $E(U_i)$ is the energy in state U_i, $\beta = 1/kT$, and T is the absolute temperature. (All these quantities refer to the system immersed in the heat bath.) The sum, of course, extends over all fundamental states of the system.

An average constructed in accordance with this principle is called a *canonical average*. The basic differences between it and the microcanonical average are: 1) All the fundamental states of the system participate in the canonical average, not just those belonging to a definite energy level, as in the microcanonical average. (This, of course, corresponds to the real difference between a system immersed in a heat bath, and hence able to change its energy, and an isolated system, whose energy remains unchanged.) 2) In the canonical average, as opposed to the microcanonical, the weights of the different fundamental states are different. The weight of any particular fundamental state depends on the corresponding energy level, so that all fundamental states belonging to the same energy level receive identical weights.

It is obvious that the principles of averaging (7) and (8) of Supplement II are valid for a system immersed in a heat bath, because they are immediate consequences of formula (5) of that supplement.

The principle of canonical averaging has many important practical advantages. In particular, canonical averages are incomparably simpler than microcanonical ones. Therefore, many authors introduce this principle from the beginning as a hypothesis and use it as the basis of all their statistical calculations. They sometimes refer to the theorem we have just proved (i.e., that microcanonical averaging for isolated systems implies canonical averaging for systems immersed in a heat bath), and state that only systems of the second type are to be considered in the sequel. However, in the majority of cases the principle of canonical averaging is introduced in a purely postulational form. It is then applied to various systems, regardless of whether they are isolated, immersed in a heat bath, or, as usually happens in practice, are in some intermediate state which only more or less approximates one of these two extreme types. In particular, this is precisely the procedure followed by Gibbs, who first introduced both these principles into statistical mechanics and who coined their universally accepted names. We may well ask whether or not such a practice is logical.

In order to answer this question we recall first of all that even the microcanonical principle was introduced as a postulate, whose arbitrariness we emphasized repeatedly. Although this choice was later given some justification by our proof of the suitability of microcanonical averages, nevertheless, as we repeatedly emphasized anew, only experiment could give final verification to this hypothesis. In particular, we examined in full detail a case where this choice was completely refuted by experiment, and had to be replaced by another choice (the introduction of the symmetric and antisymmetric principles of averaging). Gibbs and his followers in introducing the principle of canonical averaging also take this point of view that verification can come only from experiment. With the help of this principle Gibbs constructs statistical thermodynamics. The various concepts of this theory are then identified with the corresponding concepts of phenomenological thermodynamics. If, in fact, it turns out that the statistical theory is able to substantiate the fundamental formulas of phenomenological thermodynamics, which have been well verified in practice, then this is all that can be required of the theory, and the postulate which was introduced is thereby justified.

However, there are profound differences between the two principles of averaging — canonical and microcanonical. Suppose we consider an isolated system whose energy has a definite value E. This means that the system is really in one of the states belonging to the energy level E. Under these conditions, to include, in the formation of mean values of physical quantities, states whose energy levels differ from E, and in which the system cannot possibly be found, is clearly inadmissible from a theoretical point of view. The practical success of the principle of canonical averaging when applied to isolated systems, if it is valid, is not due to the inclusion of these states, but results *in spite* of their inclusion. Another characteristic difference between the two principles pertains to any type of system. While microcanonical averaging has as its basis only a very general assumption (identical weights for all admissible states), and hence is indisputably the simplest and most natural of all possible principles, the canonical principle ascribes a very special form to the weight function. We would, of course, like to know why this particular function appears and not some other one, and whether all the conclusions would remain valid if this special assumption were to be replaced by some more general one. If the canonical principle is taken as a postulate, then all these natural questions are avoided. Gibbs says of his own choice only that the canonical weight function is very convenient for performing calculations. If, however, one takes the path systematically followed in this book and 1) accepts for the treatment of isolated systems the very simple and natural microcanonical principle, 2) modifies this principle in a natural way where theory and experiment require it (the transition to the "new" statistics!), and, finally, 3) proves rigorously the canonical principle for systems immersed in a heat bath, then all the doubts and perplexing questions pointed out above disappear completely, and we get the same practical conclusions in a manner which is theoretically very satisfactory. We feel that for this purpose alone it would be worthwhile for the reader to master the fairly simple mathematical development required by this method.

Finally, one can take another point of view of the relationship between the canonical and microcanonical averages. Since we used the microcanonical principle for isolated physical systems, we were forced to find simple asymptotic expressions for the averages obtained because of the complexity of the exact expressions. (This constituted the chief problem of our method.) On the other hand, in most cases canonical averaging (as the development of the theory shows) leads to results which differ but little from those of microcanonical averaging, and one can think of the canonical principle merely as a simple and unified mathematical prescription for finding approximate values for the microcanonical averages. As such, the method of canonical averaging is completely acceptable, especially since we are forced to provide approximate expressions for the microcanonical averages anyway. However, having taken this point of view, we must consider under what conditions the canonical averages can actually serve as approximations to the microcanonical averages. In particular, we must determine the magnitude of the error in these approximate expressions.

We shall do this for the most important physical quantities, i.e., for *sum functions.*

We denote by $F(U)$ an arbitrary phase function of the system and by $<F_x>$ the microcanonical average of the quantity $F(U)$ when $E = x$ is the total energy of the system. (The average is, of course, carried out over the manifold $\mathfrak{M}_x$.) However, the canonical average of the function $F(U)$ (when the system has total energy E) is, by formula (1),

$$\ll F_E \gg = \sum_{i=1}^{\infty} F(U_i)e^{-\beta E(U_i)}\left[\sum_{x=0}^{\infty} \Omega(x)e^{-\beta x}\right]^{-1},$$

where for brevity we write $\Omega(x)$ instead of $\Omega(N, x)$, and where the value of the parameter β is determined in the well-known way from the number of particles N and the total energy E. Clearly, we can write

$$\sum_{i=1}^{\infty} F(U_i)e^{-\beta E(U_i)} = \sum_{x=0}^{\infty} e^{-\beta x} \sum_{E(U_i)=x} F(U_i)$$
$$= \sum_{x=0}^{\infty} \Omega(x)e^{-\beta x}\left[\sum_{E(U_i)=x} F(U_i)\right][\Omega(x)]^{-1} = \sum_{x=0}^{\infty} <F_x>\, \Omega(x)e^{-\beta x},$$

since $[\Omega(x)]^{-1}\sum_{E(U_i)=x} F(U_i)$ is just the microcanonical average of the function $F(U)$ when the total energy of the system has the value x.

Thus we find, for the canonical average of the function $F(U)$, the expression

$$\ll F_E \gg = \sum_{x=0}^{\infty} <F_x>\, \Omega(x)e^{-\beta x}\left[\sum_{x=0}^{\infty} \Omega(x)e^{-\beta x}\right]^{-1}.$$

This means that the canonical average $\ll F_E \gg$ of the phase function $F(U)$ is a certain weighted average of its microcanonical averages over all possible energy levels, where, to the level x, we ascribe the weight

$$p(x) = \Omega(x)e^{-\beta x}\left[\sum_{x=0}^{\infty} \Omega(x)e^{-\beta x}\right]^{-1}.$$

This weight depends, of course, on the total energy E of the system, since the parameter β depends on E. Our problem will now be to compare the canonical average $\ll F_E \gg$ of the function $F(U)$ with its microcanonical average $<F_E>$ for the same energy E.

It is easy to see, by using formulas (22) and (35) of Chapter V, that the weight function $p(x)$ [where $\Omega(x)$ stands for $\Omega(N, x)$] can be expressed approximately by the normal law

$$p(x) \approx (2\pi B)^{-\frac{1}{2}}e^{-(x-A)^2/2B},$$

where, as in Chapter V, $A = E$ as a result of our choice of the parameter β. (The two-dimensional normal law of V, §5 reduces here to a one-dimensional law, since N is constant and hence the parameter u_1 vanishes identically.) By a detailed calculation which we omit here, we find that the dispersion B is an *infinitely large quantity* of the order of E, V and N. Thus, we obtain

$$(2) \qquad \ll F_E \gg \approx (2\pi B)^{-\frac{1}{2}} \sum_{x=0}^{\infty} < F_x > e^{-(x-E)^2/2B}.$$

The weight function has its greatest value for $x = E$ and is negligibly small for values of x sufficiently far from E. Hence, in the weighted average (2) of the function $<F_x>$, only those values of $<F_x>$ receive appreciable weight for which x is sufficiently near E. This is a typical example of expressing a quantity approximately, with the help of an "integral kernel". Here, the function corresponding to the integral kernel is the weight function $p(x)$.

To illustrate the computation of the error in the equation $\ll F_E \gg \approx < F_E >$, we assume that the quantity $<F_x>$ does not change too rapidly with x near $x = E$. For example, we assume that (at least for not too large a value of $| y |$) the "Lipschitz condition"

$$| < F_{E+y} > - < F_E > | < C \, | y |$$

is satisfied for some constant $C > 0$. Then we obtain

$$| \ll F_E \gg - < F_E > | \approx (2\pi B)^{-\frac{1}{2}} \left| \sum_{x=-\infty}^{\infty} (< F_x > - < F_E >) e^{-(x-E)^2/2B} \right|$$

$$= (2\pi B)^{-\frac{1}{2}} \left| \sum_{y=-\infty}^{\infty} (< F_{E+y} > - < F_E >) e^{-y^2/2B} \right|$$

$$< C(2\pi B)^{-\frac{1}{2}} \sum_{y=-\infty}^{\infty} | y | \, e^{-y^2/2B}$$

$$\approx C(2\pi B)^{-\frac{1}{2}} \int_{-\infty}^{\infty} | y | \, e^{-y^2/2B} \, dy = CB^{\frac{1}{2}}(2\pi)^{-\frac{1}{2}} \int_{-\infty}^{\infty} | z | \, e^{-\frac{1}{2}z^2} \, dz = C^* N^{\frac{1}{2}},$$

where C^* is a constant. If the function $F(U)$ is a sum function (or the mathematical expectation of a sum function), then the quantity $<F_E>$ (as we have seen many times) will be infinitely large of order N. The last inequality shows, therefore, that in replacing the microcanonical average of a sum function by its canonical average, we introduce only a negligibly small relative error. This is precisely what we wanted to show.

Supplement IV

REDUCTION TO A ONE-DIMENSIONAL PROBLEM IN THE CASE OF COMPLETE STATISTICS

The reduction of the problem of finding estimates for structure functions to limit problems of the theory of probability was presented in V, §3. In the case of complete statistics, this reduction can be replaced by a considerably simpler one which leads to a one-dimensional limit problem. As we shall now see, such a replacement is possible because in the case of complete statistics the introduction of two parameters (α and β) can be successfully avoided by the introduction of one parameter (β). In the case of the other two statistics both parameters are necessary. (We take this opportunity to mention that the discussion of this question given in §11 of the author's article [9], although it contains no errors, is unsatisfactory since this important one-parameter property of the problem is not emphasized.)

We consider the elementary distribution law

$$(1)\qquad \mathsf{P}(z = k) = g_k e^{-\beta k}/\Phi(\beta) \qquad (k = 0, 1, \cdots),$$

where

$$\Phi(\beta) = \sum_{k=0}^{\infty} g_k e^{-\beta k} = \sum_{r=1}^{\infty} e^{-\beta \varepsilon_r}.$$

Let the random variable $R = \sum_{i=1}^{N} z_i$ be the sum of N mutually independent random variables, each of which is distributed according to the elementary law defined above. Then for a given integer E,

$$\mathsf{P}(R = E) = \sum_{k_i} \left[\prod_{i=1}^{N} \mathsf{P}(z_i = k_i)\right],$$

where the summation extends over all combinations of non-negative integers k_i satisfying the condition

$$(L)\qquad \sum_{i=1}^{N} k_i = E.$$

We abbreviate this sum by the symbol $\sum_{(L)}$. In virtue of (1),

$$(2)\qquad \mathsf{P}(R = E) = [\Phi(\beta)]^{-N} e^{-\beta E} \sum_{(L)} \prod_{i=1}^{N} g_{k_i}.$$

If a definite system of numbers k_i ($1 \leq i \leq N$) is chosen satisfying equation (L), then we may assume that the energy k_i of the ith particle is fixed. Corresponding to this energy there are g_{k_i} different fundamental states of the particle. Therefore, since a fundamental state of the system in the case of complete statistics is uniquely determined by specifying the fundamental state of each of the particles, it follows that to a given choice of the numbers k_i $(1 \leq i \leq N)$ there corresponds

$$\prod_{i=1}^{N} g_{k_i}$$

different fundamental states of the system. The total number of fundamental states of the system corresponding to the given energy level E is, therefore, equal to

$$\sum_{(L)} \prod_{i=1}^{N} g_{k_i} .$$

But this number is just $\Omega(N, E)$. Hence, formula (2) gives

$$\mathsf{P}(R = E) = [\Phi(\beta)]^{-N} e^{-\beta E} \Omega(N, E),$$

whence

$$\Omega(N, E) = [\Phi(\beta)]^{N} e^{\beta E} \mathsf{P}(R = E). \tag{3}$$

Since R is the sum of a large number N of mutually independent random variables, distributed according to the same law (which is independent of N and E), relation (3) permits us to find asymptotic expressions for $\Omega(N, E)$ with the help of well-known *one-dimensional* limit theorems of the theory of probability. It is easily seen that the asymptotic expressions obtained in this way coincide with those found in Chapter V.

Supplement V

SOME GENERAL THEOREMS OF STATISTICAL PHYSICS [10]

§1

In phenomenological thermodynamics the state of a physical system is determined by the assignment of a small number of parameters. Thus, the state of a given mass of gas, not under the influence of any external force fields, is usually determined by the assignment of its volume and temperature. Every other quantity characterizing the state of the gas is then determined as a function of these two basic quantities, and hence may be considered uniquely determined if the values of the volume and temperature are known. On the other hand, in statistical thermodynamics, to given values of the energy (or, equivalently, the temperature) and the external parameters there corresponds not one state but an uncountable set of different states of the system. (In classical mechanics this set is the entire "energy shell", and in quantum mechanics it is a linear manifold whose dimension is the degree of degeneracy of the given energy level.) Any quantity which is determined by the state of the system will generally assume different values for different states of this family, and can no longer be considered a unique function of the energy and the external parameters. Thus, there are essential differences between the phenomenological and the statistical concepts. Since, on the whole, the implications of the phenomenological theory are considered to be substantiated by experience, the statistical theory must provide an answer to the following question: How can it be that, for given values of the energy and the external parameters, a quantity which can in principle assume different values always maintains the same value experimentally, in agreement with the deductions from the phenomenological theory?

The reasons for this phenomenon were correctly guessed by the founders of the statistical theory. Boltzmann, and later Jeans, Lorentz and others pointed out repeatedly that quantities which characterize a given system in the large (and a phenomenological theory is only concerned with such quantities), though generally assuming different values in different states (which correspond to fixed values of the energy and the external parameters), nevertheless remain "almost constant". That is, for the overwhelming majority of the states of such a family, a quantity takes on values which are very near to one another. Hence, an experiment only rarely detects a value which differs significantly from a certain definite number. This number is the one predicted by the phenomenological theory as the only possible value of this quantity under the given conditions. It is also the one predicted by the statistical theory as the most probable or mean value of the quantity. This property of phenomenological quantities (which Jeans called their "normality") is explained by the fact that the value of such a quantity depends on the states of an enormous number of constituent particles. As a consequence, a mechanism which is analogous to the law of large numbers operates, and the "near-constancy" of the phenomenological quantities is thus analogous to the stability of arithmetic means in the theory of probability.

Presumably the above argument correctly characterizes the state of affairs, and, as far as I know, it has never been subject to doubts. Hence, it is all the more important to point out that the assertion contained in this argument has not only never been proved, but, so far as I know, it has never even received a precise and general mathematical formulation. Just what kind of quantities possess Jeans' "normality" and how can this property be rigorously justified? Apparently this question has not even been raised in a sufficiently general form. Until recently, one would encounter only passing remarks concerning the "normality" of some particular physical quantity. However, Fowler in his well-known treatise [11] posed and solved the problem of "normality" for a broad class of quantities, in fact for those quantities which I call sum functions. A great many of the quantities with which one is concerned in the phenomenological theory are sum functions. In my books I established the "normality" of sum functions under very broad conditions by means of a new method, based on the application of limit theorems of the theory of probability.

However, sum functions are not the only quantities of interest in a phenomenological theory. For example, the square of a sum function (an estimate of which is interesting, if only for the calculation of the dispersion) is no longer a sum function. On the other hand, it is by no means true that every quantity uniquely determined by the state of the system is of interest in the phenomenological theory. Therefore, it is necessary to define precisely the class of physical quantities which can be given a reasonable phenomenological interpretation and with respect to which one might hypothesize (and attempt to prove) the "normal" character.

In the first place, such a quantity must depend symmetrically on the constituent particles of the system. Let us discuss quantum-mechanical systems, for definiteness. As before we assume that the possible energy levels of a particle are integers. The number of linearly independent states of a particle corresponding to a given energy level r ($r = 1, 2, \cdots$) is asymptotically proportional to the volume V occupied by the system. We denote this number by Vg_r (if r does not belong to the possible energy levels of the particle, then $g_r \equiv 0$), and we denote the states themselves, enumerated in any order, by u_{rs} $(1 \leq s \leq Vg_r)$. The state of the system is known if one knows in which of the states u_{rs} each of the particles may be found. (In the case of Bose or Fermi statistics it is sufficient to know the number of particles in each of the states u_{rs}.) A physical quantity whose value is changed by a permutation of any pair of particles is meaningful in the statistical theory only for Maxwell-Boltzmann statistics. Such a quantity is never of interest in a phenomenological theory because the state of the system obtained by some permutation of a pair of particles is not, in general, a different phenomenological state.

If we denote the state of the ith particle by $u^{(i)}$, then each quantity of interest in the phenomenological theory must be a symmetric function of the states $u^{(1)}, u^{(2)}, \cdots, u^{(N)}$. However, in the great majority of cases, such a quantity will depend only on the energies of the individual particles, i.e., it will not change its value if we change some of the states $u^{(i)}$ to other states having the same energy.

In every case it is easy to show that a quantity which does not possess this property, in general does not have the normal character. Quantities of this type of the greatest physical importance are the occupation numbers a_{rs}. (a_{rs} is the number of particles in the state u_{rs}.) It is obvious that a_{rs} (in any of the three basic statistical schemes) depends symmetrically on the particles. However, when the number of particles approaches infinity, the distribution of the quantity a_{rs} tends to a certain definite limiting distribution, with a definite, positive mathematical expectation and dispersion. Hence the occupation numbers do not possess the "normal" character. On the other hand, the number N_r of particles with a given energy r $(N_r = \sum_{s=1}^{Vg_r} a_{rs})$, is a symmetric function of the states of the particles which depends only on the particle energies; and it is well-known that the numbers N_r possess the "normal" character.

Thus, we arrive naturally at the conclusion that those physical quantities which are of phenomenological interest, and which may be expected to exhibit (for a large number of particles) the "normal" character, must be symmetric functions of the energies $\varepsilon_1, \varepsilon_2, \cdots, \varepsilon_N$ of the constituent particles. The fact that this "normality" actually holds (at least in those cases where the function in question satisfies certain very broad conditions of "smoothness") amounts to a general theorem of statistical physics which I shall prove below for quantum-mechanical systems obeying any of the three statistical schemes. This theorem also holds for classical systems. However, in this case its proof is more complicated mathematically. Obviously, it can be presumed that this theorem also holds for non-homogeneous systems (i.e., systems composed of particles of several different types) if the function in question is symmetric with respect to all the particles of the same structure. In the present treatment, however, I shall not touch upon these extensions.

Every symmetric function of the energies of the particles depends only on the *numbers* of particles in the various energy levels r, i.e., it is a unique function of the numbers N_r ($r = 1, 2, \cdots$) defined above. We shall denote it by $F(N_1, N_2, \cdots, N_r, \cdots)$, or, more briefly, by $F(N_r)$. Obviously, this function can also be considered a functional, dependent on the distribution of particles among the different energy levels, i.e., dependent on the function $N(x)$—the number of particles whose energies do not exceed x. If this functional is linear, then

$$F(N_r) = \int_0^\infty \Psi(x)\, dN(x) = \sum_{r=1}^{\infty} \Psi(r)N_r = \sum_{i=1}^{N} \Psi(\varepsilon_i).$$

Obviously $F(N_r)$ is a sum function. Thus, from this point of view sum functions become the simplest group among the class of functions under consideration. [From the point of view of the pure theory of probability, the limiting behavior of such functionals has been studied carefully and with great success by von Mises [12, 13]. However (aside from the fact that he considers only finite-valued or continuously-distributed quantities, whereas we are interested in the case of a discrete but unbounded spectrum), von Mises assumed the energies of the particles to be mutually independent random variables, while the basic feature of our problem consists in taking account of their mutual dependence, which results from fixing the total energy of the system (i.e., the sum of the energies of the particles). The results and methods of von Mises can therefore not be applied to our problem.]

§2

In most cases, a function $F(N_1, N_2, \cdots, N_r, \cdots)$, which represents a quantity studied in the phenomenological theory, increases without bound as the number of particles N approaches infinity (just as, for example, do the numbers N_r). It goes without saying that by the "normality" of such a quantity we mean the requirement that, in the great majority of the states corresponding to a given energy level of the system, the function F should assume the same value, *with an error which is small compared to the value of F itself*. Thus, a sum function

$$\textstyle\sum_{i=1}^{N} \Psi(\varepsilon_i)$$

will, in general, be of order N and we shall call it normal if the errors mentioned above are of order $o(N)$. As is usually done in statistical physics, we shall always assume that the number of particles N, the energy of the system E, and the volume it occupies V all increase without bound, while maintaining constant ratios. It is under this assumption that we shall establish our asymptotic formulas. We shall also disregard the mutual potential energy of the particles in our computations, so that the energy of the system may be regarded as composed additively of the energies of its constituent particles.

As we have already noted, we must impose on the function F certain general requirements in regard to its "smoothness", i.e., the function F cannot vary too rapidly. The fact that a function which varies too rapidly cannot be "normal" can be seen by the simplest examples. In particular, the function e^{N_r} cannot be normal, as is easily seen by calculating its dispersion. The limitation on the variation of the functions of interest to us is conveniently expressed with the help of conditions of the Lipschitz type. In order to see which of these conditions is the most natural and convenient for our purposes, we consider first, as an example, a fairly general and, to a certain extent, typical class of functions

$$F(N_r) = \sum_{r=1}^{\infty} r^p N_r^q, \tag{1}$$

where $p, q \geq 1$ are positive constants. Let us put $N_r = Nl_r$ $(r = 1, 2, \cdots)$, so that $0 \leq l_r \leq 1$ and $\sum_{r=1}^{\infty} l_r = 1$. Let $N_r' = Nl_r'$ be another set of values of the numbers N_r, corresponding to the same values of N and E. Then we have

$$| F(N_r') - F(N_r) | \leq \sum_{r=1}^{\infty} r^p \,| (N_r')^q - (N_r)^q | = N^q \sum_{r=1}^{\infty} r^p \,| (l_r')^q - (l_r)^q |,$$

or, since $| (l_r')^q - (l_r)^q | \leq q \,| l_r' - l_r |$,

$$| F(N_r') - F(N_r) | \leq qN^q \sum_{r=1}^{\infty} r^p \,| l_r' - l_r |.$$

But $F(N_r) = N^q \sum_{r=1}^{\infty} r^p l_r^q$, whence

$$N^q = F(N_r)/F(l_r),$$

and we obtain

$$| F(N_r') - F(N_r) | \leq | F(N_r) | \,\varphi(l_r) \sum_{r=1}^{\infty} r^p \,| l_r' - l_r |, \tag{2}$$

where $\varphi(l_r)$ is a function of the numbers $l_1, l_2, \cdots, l_r, \cdots$.

This, then, is a Lipschitz condition satisfied by functions of the form (1). However, for our purposes we can broaden it considerably and hence prove that "normality" is possessed by a much wider class of functions. First of all we replace the quantities $| l_r' - l_r |$ by the expressions $| l_r' - l_r |^{\mu}$, where μ is an arbitrarily small positive number. Further, we replace the factor r^p by the expression $e^{\lambda r}$, where λ is an arbitrary positive number. Thus, we replace condition (2) by the much broader condition

$$| F(N_r') - F(N_r) | \leq C \,| F(N_r) | \,\varphi(l_r) \sum_{r=1}^{\infty} | l_r' - l_r |^{\mu} e^{\lambda r}. \tag{3}$$

This condition, which we impose on the function $F(N_r)$, must be understood in the following sense: A positive number μ and a positive function $\varphi(l_r)$

exist, such that, *for any value of* $\lambda > 0$, *no matter how small, inequality* (3) *is satisfied for all permissible systems of values* N_r *and* N_r', *if the constant* C *is sufficiently large.* In the following, this condition will be called the *extended Lipschitz condition.*

§3

Suppose that a system is composed of N particles and occupies a volume V. Then it is well-known that, to a given energy level E of this system, there corresponds a certain linear manifold of states, whose dimension is given by the degree of degeneracy of this eigenvalue of the Hamiltonian of the system. We shall denote this degree of degenerary by $\Omega(N, E)$. A linear basis for this manifold can be chosen in various ways [but it always consists of $\Omega(N, E)$ terms]. In particular, as was shown in detail in Chapter III, one can choose, for the elements of this basis, the states which I call fundamental. These states have the very convenient property that the numbers N_r have definite values in each of them. (In general, in a given state of the system, the numbers N_r, like all physical quantities, have only certain particular probability distributions.) In the following, we shall suppose that these fundamental states have been chosen for the basis, so that we can speak of a definite value for each number N_r in each of these states. Moreover, any function $F(N_1, N_2, \cdots, N_r, \cdots)$ of these numbers will have a definite value in each of the fundamental states.

That property of a quantity, represented by the function $F(N_r)$, which is usually called its "normality", can be precisely defined in these terms. We require the existence of a quantity $<F>$, depending on N, but not depending on the particular one of the $\Omega(N, E)$ fundamental states in which the system is found, such that the ratio

$$[F(N_r) - <F>]/<F>$$

is extremely small for the great majority of these states. More precisely: If ϵ is any positive number, and if $Q(N, E)$ is the number of those fundamental states of the energy level E in which

$$|\,[F(N_r) - <F>]/<F>\,| > \epsilon,$$

then we must have

$$Q(N, E)/\Omega(N, E) \to 0 \qquad (N \to \infty)$$

(assuming, of course, that E and V increase in proportion to N).

We now show that this normality, as just defined, is possessed by every function $F(N_r)$ which satisfies the extended Lipschitz condition defined above. In other words, *normality is an inherent property of all quantities which depend symmetrically on the energies of the particles, provided that they do not vary too rapidly*. This assertion is the content of Theorem 1, formulated below. First we must establish a lemma, with the help of which Theorem 1 can easily be proved.

All of our statements will hold for each of the three basic statistical schemes: complete statistics (Maxwell-Boltzmann, or scheme P), symmetric statistics (Bose-Einstein, or scheme S), and antisymmetric statistics (Fermi-Dirac, or scheme A). In those places where separate arguments must be used for these three schemes, we shall carry them out for each separately.

§4

We have denoted by $\Omega(N, E)$ the number of fundamental states of the system corresponding to the energy level E. Let $Q_r(M)$ be the number of these states in which $N_r = M$, i.e., in which just M particles have energy r. Let us find an expression for $Q_r(M)$ for each of the three basic statistical schemes. To this end we must consider, besides the energy spectrum of the particles of our system, another spectrum differing only in that the level r is missing ($g_r = 0$); all the other energy levels are the same (and have the same degree of degeneracy) as in the original spectrum. All quantities formed under the assumption of this second spectrum will be distinguished in the following by an asterisk (*).

In scheme P, in order to specify a fundamental state of the system it is necessary to fix the state of each particle. The number $Q_r(M)$ therefore stands for the number of such choices of states of the particles in which we have M particles with energy r. In order to realize such a choice, it is necessary first to choose from the total number N of particles, those M particles whose energies are to be equal to r. This can be done in $C(N, M)$ different ways. [We use the symbol $C(N, M)$ to denote the number of combinations of N things taken M at a time.] Furthermore, each of these M particles must be put in one of the Vg_r states which correspond to the energy level r. This, obviously, can be done in $(Vg_r)^M$ different ways. Now we must assign states to the remaining particles. There are $N - M$ remaining particles, the sum of their energies will be $E - Mr$, and their spectrum will be just the second spectrum we mentioned above. Thus, the number of choices of states for these remaining $N - M$ particles is simply equal to the number of states of a system of $N - M$ particles whose total energy is $E - Mr$ on the assumption of the second spectrum, i.e., it is equal to $\Omega^*(N - M, E - Mr)$. In the case of complete statistics, therefore, we have

$$(P) \qquad Q_r(M) = C(N, M)(Vg_r)^M\Omega^*(N - M, E - Mr).$$

Let us turn now to symmetric statistics (scheme S). Here, a state of the system is specified by assigning the numbers of particles which are in the various states. To the energy level k there correspond Vg_k linearly independent single particle states. We denote by a_{ks} the number of particles in the sth state with energy k ($k = 1, 2, \cdots ; s = 1, 2, \cdots, Vg_k ; \sum_{s=1}^{Vg_k} a_{ks} = N_k$). To each fundamental state of the system there corresponds a definite choice of occupation numbers a_{ks}, which, of course, satisfy the conditions

$$(4) \qquad \sum_{k=1}^{\infty} \sum_{s=1}^{Vg_k} a_{ks} = N; \qquad \sum_{k=1}^{\infty} k \sum_{s=1}^{Vg_k} a_{ks} = E.$$

Conversely, to each such choice there corresponds just one of the $\Omega(N, E)$ fundamental states of the system. Consequently, $Q_r(M)$ is the number of choices of occupation numbers which satisfy the relation

$$(5) \qquad N_r = \sum_{s=1}^{Vg_r} a_{rs} = M$$

and the conditions (4). But, one can choose numbers a_{rs} satisfying this relation in as many ways as there are integral solutions of the equation

$$\sum_{i=1}^{Vg_r} x_1 = M.$$

This number is $C(M + Vg_r - 1, M)$. (The easiest way to see this is to note that the number of integral solutions is equal to the coefficient of x^M in the expansion of $(1 - x)^{-Vg_r}$ in powers of x.) This choice having been made, we have to compute the number of ways that the remaining occupation numbers a_{ks} (for all $k \neq r$) can be chosen. Since this choice is bound only by the conditions

$$\sum{}^* a_{rs} = N - M; \qquad \sum{}^* ka_{ks} = E - Mr,$$

the number of possible choices is equal to $\Omega^*(N - M, E - Mr)$, and we find

$$(S) \qquad Q_r(M) = C(M + Vg_r - 1, M)\Omega^*(N - M, E - Mr).$$

[The asterisk means that the term $k = r$ is missing from the sum.]

Let us turn finally to the antisymmetric case (scheme A). Here there is only one difference from scheme S: Instead of arbitrary solutions of equation (5), we must count only those for which $a_{rs} \leq 1$ ($s = 1, 2, \cdots, Vg_r$), which, obviously, gives $C(Vg_r, M)$ solutions. Thus, we find

$$(A) \qquad Q_r(M) = C(Vg_r, M)\Omega^*(N - M, E - Mr).$$

Hence, our first problem is solved. To avoid misunderstanding, one must keep in mind that the expression $\Omega^*(N - M, E - Mr)$ in these formulas has different values for the three different statistical schemes.

§5

If B stands for some condition, depending on the numbers N_r, then in general this condition will be satisfied for some of the $\Omega(N, E)$ fundamental states of the system belonging to the energy level E, and will not be satisfied for the others. For brevity, let us denote by $\mathsf{P}(B)$ the fraction of these fundamental states for which condition B is satisfied. We call $\mathsf{P}(B)$ the *probability* of condition B. In particular,

$$\mathsf{P}(N_r = M) = Q_r(M)/\Omega(N, E),$$

where $Q_r(M)$ is the number we defined above. Thus, we have the following formulas for this probability:

$$(6)\quad \mathsf{P}(N_r = M) = \begin{cases} C(N, M)(Vg_r)^M\Omega^*(N - M, E - Mr)/\Omega(N, E), & (P) \\ C(M + Vg_r - 1, M)\Omega^*(N - M, E - Mr)/\Omega(N, E), & (S) \\ C(Vg_r, M)\Omega^*(N - M, E - Mr)/\Omega(N, E). & (A) \end{cases}$$

Now let α and β be two real parameters whose values are such that the following double series will converge:

$$\Phi(\alpha, \beta) = \sum_{p=0}^{\infty} \sum_{q=0}^{\infty} e^{-\alpha p - \beta q}\Omega(p, q)/C(p),$$

where $C(p) = p!$ in the case of complete statistics and $C(p) = 1$ in the other two cases. Then the following basic formula [Chapter V, §3, equation (22)] is valid:

$$(7)\qquad \Omega(p, q) = C(p)\Phi(\alpha, \beta)e^{\alpha p + \beta q}P(p, q),$$

where $P(p, q)$ is a certain two-dimensional probability distribution. We shall have more to say about this quantity later on. We have the following expressions for the function $\Phi(\alpha, \beta)$ (see p. 157):

$$(8)\qquad \ln \Phi(\alpha, \beta) = \begin{cases} \sum_{k=1}^{\infty} Vg_k\, e^{-(\alpha+\beta k)}, & (P) \\ -\sum_{k=1}^{\infty} Vg_k \ln [1 - e^{-(\alpha+\beta k)}], & (S) \\ \sum_{k=1}^{\infty} Vg_k \ln [1 + e^{-(\alpha+\beta k)}]. & (A) \end{cases}$$

If we turn to our "second" system and put (in accordance with our system of notation)

$$\Phi^*(\alpha, \beta) = \sum_{p=0}^{\infty} \sum_{q=0}^{\infty} e^{-\alpha p - \beta q}\Omega^*(p, q)/C(p),$$

then $\ln \Phi^*(\alpha, \beta)$ is given by the same sums (8), except that the term $k = r$ is missing; whence

$$\ln[\Phi^*(\alpha,\beta)/\Phi(\alpha,\beta)] = \begin{cases} -Vg_r e^{-(\alpha+\beta r)}, & (P) \\ Vg_r \ln[1 - e^{-(\alpha+\beta r)}], & (S) \\ -Vg_r \ln[1 + e^{-(\alpha+\beta r)}]. & (A) \end{cases}$$

Applying formula (7) to the numerator and denominator of the fraction $\Omega^*(N - M, E - Mr)/\Omega(N, E)$, we find

$$\begin{aligned} \Omega^*(N - M, E - Mr)/\Omega(N, E) \\ &= [C(N-M)/C(N)][\Phi^*(\alpha,\beta)/\Phi(\alpha,\beta)]e^{-M(\alpha+\beta r)} \\ &\qquad \cdot P^*(N - M, E - Mr)/P(N, E) \\ &= [C(N-M)/C(N)]e^{-M(\alpha+\beta r)}R_M \begin{cases} \exp[-Vg_r e^{-(\alpha+\beta r)}], & (P) \\ [1 - e^{-(\alpha+\beta r)}]^{Vg_r}, & (S) \\ [1 + e^{-(\alpha+\beta r)}]^{-Vg_r}, & (A) \end{cases} \end{aligned}$$

where

$$R_M = P^*(N - M, E - Mr)/P(N, E).$$

We now substitute this expression in formula (6). For brevity we introduce the "index of symmetry" σ of the given system [$\sigma = 0$ for (P), $\sigma = 1$ for (S), $\sigma = -1$ for (A)], and we put

$$(e^{\alpha+\beta r} - \sigma)^{-1} = T_r \qquad [(P), (S), (A)]\cdot$$

Then we easily obtain

$$(9) \quad \mathsf{P}(N_r = M) = \begin{cases} [e^{-Vg_r T_r}(Vg_r T_r)^M/M!]R_M, & (P) \\ C(M + Vg_r - 1, M) \\ \qquad \cdot (T_r + 1)^{-Vg_r}[T_r/(T_r + 1)]^M R_M, & (S) \\ C(Vg_r, M)T_r^M(1 - T_r)^{Vg_r - M}R_M. & (A) \end{cases}$$

In these expressions it is interesting to note that in all three cases the factors preceding R_M represent a very simple probability distribution, corresponding to a certain integral-valued random variable. In all three cases we shall call it the "basic distribution".

§6

First of all, let us estimate the factor

$$R_M = P^*(N - M, E - Mr)/P(N, E).$$

The denominator of this expression (which is independent of M) can be obtained from formula (37) of Chapter V, §5. Thus,

(10) $$P(N, E) = d/2\pi V\delta^{\frac{1}{2}} + O(V^{-2}),$$

where d and δ are positive constants.

The estimate of the numerator is much more complicated. However, for our purposes it is sufficient to consider only values of M which satisfy the inequality

(11) $$|M - Vg_rT_r| > T_r^{\epsilon}(Vg_r)^{1-\epsilon},$$

where ϵ is some constant ($0 < \epsilon < \frac{1}{4}$). Moreover, we require only a very crude estimate which can be derived easily by elementary means.

By definition, $P^*(N - M, E - Mr)$ is the probability distribution of a certain integral-valued random vector (S^*, T^*). Hence

$$P^*(N - M, E - Mr) = \mathsf{P}(S^* = N - M, T^* = E - Mr)$$
$$\leq \mathsf{P}(S^* = N - M) \leq \sum_{(11)} \mathsf{P}(S^* = N - M),$$

where the summation extends over all values of M which satisfy condition (11). It follows that

(12) $$P^*(N - M, E - Mr) \leq \mathsf{P}\{|S^* - N + Vg_r T_r| > T_r^{\epsilon}(Vg_r)^{1-\epsilon}\}.$$

Until now the values of the parameters α and β have been arbitrary. Now we shall choose them in the usual way, so that the relations

$$N + \partial \ln \Phi/\partial\alpha = 0; \qquad E + \partial \ln \Phi/\partial\beta = 0$$

are satisfied. (It is well-known that such a choice can always be made uniquely, and that the resulting values of α and β are constant, i.e., do not depend on N, E and V.) It is known (p. 139) that the mathematical expectation of the quantity S^* is

$$\mathsf{E}S^* = -\partial \ln \Phi^*/\partial\alpha = -\partial \ln \Phi/\partial\alpha - Vg_r T_r$$
$$= N - Vg_r T_r.$$

Using this, and applying the Chebyshev inequality to inequality (12), we obtain

$$P^*(N - M, E - Mr) \leq \mathsf{P}\{|S^* - \mathsf{E}S^*| > T_r^{\epsilon}(Vg_r)^{1-\epsilon}\}$$
$$\leq \mathsf{D}S^*(Vg_r/T_r)^{2\epsilon}(Vg_r)^{-2},$$

where $\mathsf{D}S^*$ is the dispersion of the quantity S^*. But S^* is the sum of V mutually independent random variables which have the same *constant* (i.e., not depending on N, E and V) probability distribution. If we denote the dispersion of this elementary distribution by b^*, it follows that

$$\mathsf{D}S^* = Vb^*,$$

and we find

$$P^*(N - M, E - Mr) \leq b^*(Vg_r/T_r)^{2\epsilon}V^{-1}g_r^{-2}.$$

We shall be satisfied with this very crude estimate [which holds for all values of M satisfying condition (11)]. Combining it with estimate (10) for $P(N, E)$, we find, for all such values of M and for all three statistics,

$$(13) \quad R_M = P^*(N - M, E - Mr)/P(N, E) \leq c(Vg_r/T_r)^{2\epsilon}g_r^{-2}.$$

(Here c is some positive number, independent of V and r. A simple computation, which we omit, shows that

$$b^* = V^{-1}\partial^2 \ln \Phi^*/\partial\alpha^2 \leq V^{-1}\partial^2 \ln \Phi/\partial\alpha^2,$$

where the right side is independent of V and r.)

It is always assumed, of course, that the number r is one of the possible energy levels of a particle, i.e., that $g_r > 0$.

§7

Now let us denote by P_0, E_0 and D_0 the probability, mathematical expectation and dispersion, respectively, of the integral-valued random variable corresponding to the basic distribution defined by the factor preceding R_M in each of the three formulas (9). Again applying the Chebyshev inequality, we find, for all three statistics,

$$\begin{aligned}
\mathsf{P}\{ | N_r - Vg_rT_r | > T_r^{\epsilon}(Vg_r)^{1-\epsilon}\} &= \textstyle\sum_{(11)} P_0(M)R_M \\
&\leq c(Vg_r/T_r)^{2\epsilon}g_r^{-2}\textstyle\sum_{(11)} P_0(M) \\
(14) \qquad &= c(Vg_r/T_r)^{2\epsilon}g_r^{-2}P_0\{ | N_r - Vg_rT_r | > T_r^{\epsilon}(Vg_r)^{1-\epsilon}\} \\
&\leq c(Vg_r/T_r)^{4\epsilon}V^{-2}g_r^{-4}E_0\{[N_r - Vg_rT_r]^2\} \\
&= c(Vg_r/T_r)^{4\epsilon}V^{-2}g_r^{-4}D_0(N_r).
\end{aligned}$$

The above follows since each of the three basic distributions has (as is shown by an elementary computation) the mathematical expectation Vg_rT_r, and hence in all three cases

$$E_0\{[N_r - Vg_rT_r]^2\} = D_0(N_r).$$

But it is also easily shown that in all three cases

$$D_0(N_r) = Vg_rT_r(1 + \sigma T_r),$$

where σ is the index of symmetry. Therefore, inequality (14) gives

$$\mathsf{P}\{ | N_r - Vg_rT_r | > T_r^{\epsilon}(Vg_r)^{1-\epsilon}\} \leq c(T_r/Vg_r)^{1-4\epsilon}g_r^{-2}(1 + \sigma T_r),$$

or, since $g_r \geq 1$,

$$\mathsf{P}\{\,|N_r - Vg_rT_r| > T_r^{\epsilon}(Vg_r)^{1-\epsilon}\} \leq c(T_r/V)^{1-4\epsilon}(1 + \sigma T_r).$$

This is our estimate of the probability of the inequality

$$(15) \qquad |N_r - Vg_rT_r| > T_r^{\epsilon}(Vg_r)^{1-\epsilon}$$

for a fixed value of r. The probability that inequality (15) holds *for at least one value of* r $(1 \leq r < \infty)$, does not exceed the sum of the probabilities just estimated, i.e., it does not exceed the quantity

$$cV^{-(1-4\epsilon)}\sum_{r=1}^{\infty} T_r^{1-4\epsilon}(1 + \sigma T_r) = c'V^{-(1-4\epsilon)}$$

(where c' is a positive constant). Thus, we have proved the following proposition:

LEMMA. *With a probability greater than*

$$1 - c'V^{-(1-4\epsilon)}$$

we can assert that for any r $(1 \leq r < \infty)$,

$$(16) \qquad |N_r - Vg_rT_r| \leq T_r^{\epsilon}(Vg_r)^{1-\epsilon},$$

where ϵ *is any positive number in the interval* $0 < \epsilon < \frac{1}{4}$, *and* c' *is some positive constant.* According to our definition of probability this means that the whole set of inequalities (16) will hold for all fundamental states of the energy level E with the exception of no more than $c'V^{-(1-4\epsilon)}$ of the states. (Thus, the fraction of the fundamental states represented by the exceptional ones tends to zero with increasing N, E and V.)

§8

THEOREM 1. *A function* $F(N_1, N_2, \cdots, N_r, \cdots)$, *which satisfies the extended Lipschitz condition, is normal.*

Let us put

$$N_r/N = l_r, \qquad (V/N)g_rT_r = \lambda_r \qquad (r = 1, 2, \cdots),$$

so that the numbers λ_r can be considered constants. Then the quantity

$$<F> = F(N\lambda_1, N\lambda_2, \cdots, N\lambda_r, \cdots)$$

will depend on N (and hence on E and V) but not on the particular state of the system (and, in particular, it will not depend on the numbers N_r). Since the function F by hypothesis satisfies the extended Lipschitz condition, we have, by (3), for any $\alpha > 0$,

$$(17) \qquad |F(N_r) - <F>| \leq C<F>\varphi(\lambda_r)\sum_{r=1}^{\infty} |l_r - \lambda_r|^{\mu}e^{\lambda r},$$

where $\varphi(\lambda_r) = \varphi(\lambda_1, \lambda_2, \cdots)$ depends only on the numbers λ_r and is thus constant.

Now we decompose the set of all $\Omega(N, E)$ fundamental states of the system for the energy level E into two classes K_1 and K_2, putting those states into class K_1 for which inequality (16) holds for some r. The remaining states are put into class K_2. We denote by $\Omega_1(N, E)$ and $\Omega_2(N, E)$ the numbers of states in these two classes, respectively. By our lemma,

$$\Omega_2(N, E)/\Omega(N, E) \leq c'V^{-(1-4\epsilon)} \to 0 \qquad (N \to \infty).$$

On the other hand, for each state of class K_1, we have

$$| l_r - \lambda_r | = N^{-1} | N_r - Vg_rT_r | \leq c_1 g_r(T_r/V)^\epsilon \leq c_2 g_r e^{-\epsilon\beta r} V^{-\epsilon},$$

where c_1 and c_2 are positive constants; whence

$$| l_r - \lambda_r |^\mu \leq c_2^\mu g_r^\mu e^{-\epsilon\mu\beta r} V^{-\epsilon\mu}.$$

Thus, for any state of the class K_1, inequality (17) gives (if we remember that $\varphi(\lambda_r)$ is constant)

$$| F(N_r) - <F> | \leq c_3 V^{-\epsilon\mu} <F> \sum_{r=1}^{\infty} g_r^\mu e^{(\lambda - \epsilon\mu\beta)r}.$$

If we choose $\lambda < \epsilon\mu\beta$, the series on the right converges and its sum is some constant; therefore,

$$(18) \qquad | F(N_r) - <F> | \leq c_4 V^{-\epsilon\mu} <F>.$$

But this establishes the normality of the function $F(N_r)$ since inequality (18) holds for all states of class K_1, which comprises the overwhelming majority of the $\Omega(N, E)$ fundamental states with energy E.

Moreover, we have obtained a simple method of finding the quantity $<F>$ which is representative of the values of F in this majority of states. To obtain $<F>$ we must substitute in the expression of the function F not the numbers N_r but their microcanonical averages Vg_rT_r $(r = 1, 2, \cdots)$. The number

$$<F> = F(Vg_1T_1, Vg_2T_2, \cdots)$$

will also be the value predicted by the statistical theory for the quantity expressed by the function $F(N_1, N_2, \cdots)$ for given values of N, E and V.

§9

It is well-known that the microcanonical averaging of a quantity, for given values of N, E and V, refers to the simple arithmetic average of its values for the $\Omega(N, E)$ corresponding fundamental states. In other words, in microcanonical averaging, all accessible states have the same *weight*. If we normalize these weights so that their sum over all accessible states is unity, then we call the weights *probabilities* of the corresponding states. Thus, one can say that in microcanonical averaging, all accessible states have the same probability $1/\Omega$. [Here, and in the following, we write Ω instead of $\Omega(N, E)$ for brevity.]

In statistical physics after the work of Gibbs another principle of averaging, called canonical averaging, became widely used. In canonical averaging, as distinct from microcanonical, all fundamental states consistent with the given values of N and V are assigned positive probabilities, regardless of the energy of the system. Of course, these probabilities can no longer all be equal to one another (since there are infinitely many such states). In canonical averaging each state with energy E is assigned the probability

$$ce^{-\beta E},$$

where β is a positive constant and where c is a coefficient chosen so that the sum of the probabilities of all fundamental states will be 1.

Canonical averaging is very widely used, chiefly because it is much easier to calculate averages and their properties using the canonical distribution than it is to calculate using the microcanonical distribution. But, with respect to the theoretical principles of statistical physics, the canonical method of averaging at first sight always gives the impression of being rather arbitrary. Moreover, in the formation of mean values the canonical method unnecessarily brings in states of the system which in general cannot exist for a given energy E, i.e., states with other values of the energy. (In the microcanonical method the probabilities of all such states is simply 0.) Another disadvantage is that canonical averaging makes use of a very special distribution function which cannot be justified *a priori* except in that it is very convenient for computational purposes. (The microcanonical average, on the other hand, is based on the very simple distribution in which all accessible states are given the same probability.)

There are several different ways to overcome these difficulties. For example, one can show that in a very broad class of problems, including most of the ones frequently met in practice, canonical averages are practically identical with microcanonical averages. Hence, with complete justification we can consider canonical averages as convenient mathematical approximations to microcanonical averages. (See Supplement III.) However, the difficulties under discussion are usually approached from another point of view which is also completely satisfactory. This approach is based on a mathematical theorem proved by Gibbs, under special assumptions, and later generalized considerably by other authors. This theorem asserts that *if a system is subject to a microcanonical distribution, then a small part of that system is necessarily subject to a canonical distribution.* In particular, if we are dealing with a homogeneous system composed of N particles of identical structure, then it follows from the microcanonical distribution of this system that the individual particles are distributed canonically. More accurately, as $N \to \infty$, the probability that a particle will have the energy r approaches the quantity

$$(V/N)g_rT_r = (V/N)g_r[e^{\alpha+\beta r} - \sigma]^{-1}$$

which is asymptotically proportional to $g_re^{-\beta r}$ for large r. (Since Vg_r states of a particle correspond to the energy level r, it follows that the probability of finding a particle in any one of these states is asymptotically proportional to T_r.)

Thus, we no longer consider the original system to be isolated, but assume it to be in a state of free energetic contact with its very large surroundings ("heat bath"). We then take the "original system plus heat bath" as a new system, and consider the original system as a small part of this new system. We think of this new composite system as isolated, so that its energy is constant, and hence we assume that it is distributed microcanonically. The original system being just a small part will now be distributed canonically in virtue of Gibbs' theorem.

With this approach to canonical averaging the appearance of the canonical distribution does not represent an arbitrary assumption, but is a necessary consequence of the original hypothesis. The assumption that the total system is distributed microcanonically remains as the only hypothesis. We shall see that by using Theorem 1 (proved above) we can to a very great extent remove even this hypothetical element, and extend Gibbs' theorem to the case of a system whose distribution is to a large degree arbitrary.

§10

A system of energy E and volume V consisting of N particles of identical structure is distributed microcanonically if each of the $\Omega(N, E) = \Omega$ accessible fundamental states has the probability $1/\Omega$. Let us assume now that this probability can be different for different states, but that it is a symmetric function of the particle energies. (In this case antisymmetry would be in violent contradiction to our basic assumption about the equivalence of the particles.) In other words, this probability is a function of the numbers $N_1, N_2, \cdots, N_r, \cdots$, and can be conveniently represented in the form

$$F(N_1, N_2, \cdots, N_r, \cdots)/\Omega. \tag{19}$$

For the microcanonical distribution $F \equiv 1$.

THEOREM 2. *If the probabilities of the different states of the system, consistent with the given values of N, V and E are defined by* (19), *where the function $F(N_1, N_2, \cdots, N_r, \cdots)$ is bounded and satisfies the extended Lipschitz condition, then, as $N \to \infty$, the probability that an individual particle will have energy r approaches*

$$(V/N)g_rT_r = (V/N)g_r[e^{\alpha+\beta r} - \sigma]^{-1}.$$

Let us denote by K the set of all choices of the numbers N_r which satisfy the relations

$$(K) \qquad \textstyle\sum_{r=1}^{\infty} N_r = N; \qquad \sum_{r=1}^{\infty} rN_r = E,$$

and let us put

$$F(Vg_1T_1, Vg_2T_2, \cdots, Vg_rT_r, \cdots) = <F>.$$

Let K_1 denote that part of K for which

$$(20) \qquad | F - <F> | < c_4V^{-\epsilon\mu}<F>,$$

and let K_2 denote the remaining part. As we showed in the proof of Theorem 1, the number of states corresponding to K_2 does not exceed $\Omega\, c'/V^{1-4\epsilon}$. Let us denote by $\Delta = \Delta(N_1, N_2, \cdots, N_r, \cdots)$ the number of different states of the system corresponding to a given choice of the numbers N_r. Then

$$\mathsf{E}N_r = \Omega^{-1}\textstyle\sum_{(K)} N_r\, F(N_1, N_2, \cdots, N_r, \cdots)\Delta(N_1, \cdots, N_r, \cdots),$$

where the summation extends over all sets K of choices of the numbers N_r. Whence,

$$(21) \qquad \mathsf{E}N_r = <F>\Omega^{-1}\textstyle\sum_{(K)} N_r\Delta + \Omega^{-1}\sum_{(K_1)} N_r[F - <F>]\Delta$$
$$+ \Omega^{-1}\textstyle\sum_{(K_2)} N_r[F - <F>]\Delta.$$

Since, for any of these choices of the numbers N_r, we have

$$\textstyle\sum_{r=1}^{\infty} N_r = N,$$

summing (21) over all values of r yields

$$(22) \qquad N = <F>N + N\Omega^{-1}\textstyle\sum_{(K_1)} [F - <F>]\Delta$$
$$+ N\Omega^{-1}\textstyle\sum_{(K_2)} [F - <F>]\Delta.$$

But in K_1 inequality (20) is satisfied; therefore, remembering the boundedness of the function F, we obtain

$$| N\Omega^{-1}\textstyle\sum_{(K_1)} [F - <F>]\Delta | \leq c_4<F>NV^{-\epsilon\mu} \leq c_5\, N^{1-\mu\epsilon}.$$

On the other hand, $\sum_{(K_2)} \Delta \leq \Omega\, c'/V^{1-4\epsilon}$, and consequently, if we denote the upper bound of the function F by M, we have

$$| N\Omega^{-1}\textstyle\sum_{(K_2)} [F - <F>]\Delta | \leq N\Omega^{-1}(2M\Omega\, c'/V^{1-4\epsilon}) \leq c_6\, N^{4\epsilon}.$$

Thus, relation (22) gives

$$| N(1 - <F>) | \leq c_5\, N^{1-\mu\epsilon} + c_6\, N^{4\epsilon} = o(N),$$

and consequently $<F> \to 1$ as $N \to \infty$. Since, on the other hand, the sum

$$\Omega^{-1} \sum_{(K)} N_r \, \Delta$$

is obviously the microcanonical average $\mathsf{E}_\mu N_r$ of the number N_r, relation (21) gives

$$| \mathsf{E} N_r - [1 + o(1)] \mathsf{E}_\mu N_r | < c_5 N^{1-\mu\epsilon} + c_6 N^{4\epsilon}.$$

Using the well-known formula of quantum statistics

$$\mathsf{E}_\mu N_r = V g_r T_r + o(N),$$

we find

$$\mathsf{E} N_r = V g_r T_r + o(N).$$

But the probability that a particle have the energy r is obviously equal to

$$P_N(r) = N^{-1} \mathsf{E} N_r .$$

Therefore, we find

$$P_N(r) = (V/N) g_r T_r + o(1),$$

and Theorem 2 is proved.

Supplement VI

SYMMETRIC FUNCTIONS ON MULTI-DIMENSIONAL SURFACES [14]

§1. Introduction

The general properties of symmetric functions on multi-dimensional surfaces are very important for the mathematical foundation of the principles of classical statistical physics. If the surface is a "surface of constant energy" in the phase space of a physical system, then every point of the surface describes a definite state of the system, consistent with the given total energy. In classical physics, any physical quantity associated with a system is uniquely determined by the state of that system and, hence, for a fixed total energy, is a function of position on the corresponding "energy shell".

Obviously, one can consider that every *macroscopic* quantity, which is expressed as a function of the Hamiltonian coordinates of the particles of the system, will (in the case of a homogeneous system) always be symmetric with respect to these particles. (From the mathematical point of view, it would perhaps be expedient to define a macroscopic quantity directly as any function of the particle coordinates which is symmetric with respect to these particles.) Therefore, from the physical point of view, the general properties of such symmetric functions will express the general properties of macroscopic quantities for a system with a fixed total energy.

Among these properties, the "near-constancy" of such functions has the most important physical implications: When the system consists of a large number of particles, such a function, as a rule, assumes values which are extremely close to one another for the great majority of points on the energy shell. It is clear that all these values will be very close to the microcanonical average of the function over the given energy shell. This fact is especially important because it makes the introduction of an "ergodic" theorem or hypothesis unnecessary for the study of such systems. In fact, if for the great majority of points on the energy shell the values of the function in question are very near to its average over the shell, then it is immediately obvious that the time averages along the majority of the system trajectories will practically coincide with this average over the shell. However, establishing this nearness of "time-" and "phase-" averages is precisely the goal of any "ergodic" theory.

Of course, this "near-constancy" of macroscopic quantities over an energy shell was well-known to the founders of statistical physics. One finds clear hints of this property in Jeans, Lorentz and many other authors. It is rather surprising that, up to the present, apparently no one has attempted either to prove this property of symmetric functions under some fairly general hypotheses or to give a clear definition of the corresponding mathematical problem. In Fowler's *Statistical Mechanics* [11] this "near-constancy" is established for the special class of quantities which I have called sum functions. In my two books on statistical mechanics, I gave a proof for this same class of quantities by a new method, namely by reducing the problem to one involving the local limit theorems of the theory of probability. Later (see Supplement V), I used this method to prove the "near-constancy" for very general symmetric functions in the case of the simplest type of quantum systems. The corresponding problem in classical physics is much more complicated. A quite general solution is given for the first time in this paper. The only condition necessary for the "near-constancy" of a symmetric function over an energy shell is that the function be continuous, where continuity is understood in a certain very broad sense.

The basic result mentioned above is contained in §9. The intervening sections are concerned partly with preliminary results and partly with digressions, in which certain problems are solved which have no direct connection with the main development of the paper. Thus, in §§5 and 7, we obtain an accurate solution to the problem of finding the limiting distributions of the different terms of a sequence obtained by arranging the energies of the particles comprising the system in order of increasing magnitude. This problem, no doubt, is of interest both in multi-dimensional geometry and in physics.

Finally, in §10, our basic result is applied to the generalization of a theorem of Gibbs. This generalization seems quite significant to me. The discussion concerns the theorem that a small part of a system, under very broad hypotheses, is distributed canonically in its own phase space when the system as a whole is microcanonically distributed on an energy shell. This theorem is usually (and justly) considered the most important theoretical argument in favor of the widely used canonical method of averaging. The only remaining hypothesis is the assumption about the microcanonical distribution of the "large" system. In §10 I show that this hypothesis can be changed to a much broader one: It is sufficient to assume that the distribution of the system over the energy shell is symmetric with respect to the particles (and that certain natural and general requirements of boundedness and continuity are satisfied). In fact, according to the results of §9, any continuous symmetric distribution of the system over the energy shell is "almost equal" to a microcanonical distribution.

§2. Preliminary formulas

In this section we collect several well-known formulas of classical statistical mechanics which will be used later.

Let us consider mechanical systems composed of a very large number N of particles of identical structure. We shall denote the energy of the system by E and the energy of the ith particle by e_i. We shall assume that

$$E = \sum_{i=1}^{N} e_i .$$

The set of Hamiltonian variables of the ith particle will be called briefly (q_i, p_i). Any physical quantity associated with the ith particle is a function of these variables. In particular, the energy e_i equals $e(q_i, p_i)$, where the form of the function $e(q_i, p_i)$ does not depend on i. (This just expresses the fact that the particles have identical structure.) The Euclidean space whose Cartesian coordinates are the Hamiltonian variables of a particular particle is called the phase space of that particle. We shall denote by $\mathbf{v}(x)$ the volume of that part of this phase space in which $e(q_i, p_i) < x$. We assume that the zero value of the energy is chosen so that $\mathbf{v}(0) = 0$.

The function $\omega(x) = \mathbf{v}'(x)$, which plays a most important role in our work, will be called the structure function of the particle. The hypersurface $e(q_i, p_i) = x$ of the phase space of the ith particle will be called the "surface of constant energy x". Denoting by $d\sigma$ the "surface element" of this surface, we have the well-known result

$$\omega(x) = \int d\sigma/\text{grad } e(q_i, p_i),$$

where the integration is carried out over the entire surface $e(q_i, p_i) = x$.

We introduce analogous concepts for the system composed of N particles. The Cartesian coordinates of the phase space of this system are the Hamiltonian variables (q_i, p_i) of all the particles $(i = 1, \cdots, N)$. The volume of this space in which

$$\sum_{i=1}^{N} e(q_i, p_i) < x$$

is denoted by $\mathbf{V}_N(x)$, and the function

$$\Omega_N(x) = \mathbf{V}_N'(x)$$

is called the structure function of the system. If $d\Sigma_N$ is the "surface element" of the energy shell

$$\sum_{i=1}^{N} e(q_i, p_i) = x, \tag{1}$$

then

$$\Omega_N(x) = \int d\Sigma_N/\text{grad } E,$$

where the integration is carried out over the entire shell (1) and where

$$E = \sum_{i=1}^{N} e(q_i, p_i)$$

is the total energy of the system.

The functions $\mathbf{V}_N(x)$ and $\Omega_N(x)$ for the system are related to the corresponding functions for one particle by simple composition rules. The relation

$$\mathbf{V}_N(x) = \int_{z_1+\cdots+z_N<x}^{(N)} \prod_{i=1}^{N} \{\omega(z_i)\, dz_i\}$$

$$= \int^{(N-1)} \left[\prod_{i=1}^{N-1} \{\omega(z_i)\, dz_i\}\right] \int_0^{x-(z_1+\cdots+z_{N-1})} \omega(z_N)\, dz_N$$

is almost obvious, and a differentiation with respect to x leads to the composition rule for the structure functions

$$\Omega_N(x) = \int^{(N-1)} \left[\prod_{i=1}^{N-1} \{\omega(z_i)\, dz_i\}\right] \omega\left(x - \sum_{i=1}^{N-1} z_i\right), \tag{2}$$

where the integral is extended over the entire $(N - 1)$-dimensional space.

Formally, this rule is completely analogous to the composition rule of probability densities for the sum of mutually independent random variables. This circumstance makes it possible to estimate the quantity $\Omega_N(x)$ by using the local limit theorems of the theory of probability, as we shall see in detail below.

The physical state of a system is determined by the location of a point in its phase space. Every physical quantity which has a definite value in each state of the system is therefore some function $f(P)$ of the point P in its phase space. The quantity $<f>$, defined by

$$<f> = [\Omega_N(x)]^{-1} \int f(P)\, d\Sigma_N/\text{grad } E,$$

will be called the *mean value* of $f(P)$ for a system with total energy x. The integration is extended over the constant energy surface $E = x$. This is usually called the *microcanonical* principle of averaging. In particular, this allows us to define the *probabilities* of various relations among the Hamiltonian variables of the system. In fact, for any such relation A we can put $f(P) = 1$ if relation A is satisfied at the point P, and $f(P) = 0$ otherwise. Then the probability of the event A (for $E = x$) is defined as the microcanonical average of $f(P)$ over the surface $E = x$.

The following relation ([1; p. 37]) is often useful for computing microcanonical averages:

$$<f> = [\Omega_N(x)]^{-1} (d/dx) \int_{E<x} f(P)\, d\mathbf{V}, \tag{3}$$

where $d\mathbf{V}$ stands for the volume element of the phase space of the system and the integral extends over that part of this space in which $E < x$. Let us consider, as an example, the important case in which $f(P)$ depends only on the particle energies:

$$f(P) = \Phi(e_1, \cdots, e_N).$$

Then we have

$$\begin{aligned}\int_{E<x} f(P)\, d\mathbf{V} &= \int_{e_1+\cdots+e_N<x} \Phi(e_1, \cdots, e_N)\, d\mathbf{V}\\ &= \int_{z_1+\cdots+z_N<x}^{(N)} \Phi(z_1, \cdots, z_N) \prod_{i=1}^{N} \{\omega(z_i)\, dz_i\}\\ &= \int^{(N-1)} \left[\prod_{i=1}^{N-1} \{\omega(z_i)\, dz_i\}\right]\\ &\qquad \cdot \int_0^{x-(z_1+\cdots+z_{N-1})} \Phi(z_1, \cdots, z_N)\omega(z_N)\, dz_N\,.\end{aligned}$$

By (3), we then get

$$(4)\quad <f> = [\Omega_N(x)]^{-1} \int^{(N-1)} \Phi\left(z_1, \cdots, z_{N-1}, x - \sum_{i=1}^{N-1} z_i\right) \cdot \left[\prod_{i=1}^{N-1} \{\omega(z_i)\, dz_i\}\right] \omega\left(x - \sum_{i=1}^{N-1} z_i\right).$$

This formula will be quite useful to us later. In the special case for which $f(P) = \varphi(e_1)$ depends on the energy of only one (say the first) particle, (4) gives

$$<f> = <\varphi(e_1)> = [\Omega_N(x)]^{-1} \int_0^{\infty} \varphi(z_1)\omega(z_1)\, dz_1 \cdot \int^{(N-2)} \left[\prod_{i=2}^{N-1} \{\omega(z_i)\, dz_i\}\right] \omega\left(x - \sum_{i=1}^{N-1} z_i\right).$$

According to (2), the inner integral on the right side equals $\Omega_{N-1}(x - z_1)$, and we find

$$(5)\qquad <\varphi(e_1)> = [\Omega_N(x)]^{-1} \int_0^{\infty} \varphi(z)\omega(z)\Omega_{N-1}(x - z)\, dz.$$

Another important application of formula (3) is to express the distribution of an individual particle in its phase space. For definiteness, let us speak of the first particle. Let us denote its phase space by Γ_1 and let us determine the probability that the Hamiltonian variables (q_1, p_1) of this particle belong to a certain region Δ of the space Γ_1. This probability will be the microcanonical average of the function $f(P)$ (where P, as always, denotes a point of the surface $E = x$) defined in the following manner: $f(P) = 1$ if the point P is such that $(q_1, p_1) \in \Delta$, and $f(P) = 0$ otherwise. Clearly, since $f(P)$ depends only on the Hamiltonian variables (q_1, p_1) of the first particle

$$\int_{E<x} f(P)\, d\mathbf{V} = \int_{\Gamma_1} f(P)\, d\mathbf{v} \int_{E^*<x-e_1} d\mathbf{V}^*.$$

E^* and $d\mathbf{V}^*$ denote, respectively, the energy and volume element of the phase space for the rest of the system (i.e., the set of all particles other than the first), and e_1 is the energy of the first particle at the given point of the space Γ_1. Obviously,

$$\int_{E^*<x-e_1} d\mathbf{V}^* = \mathbf{V}_{N-1}(x - e_1),$$

so that

$$\int_{E<x} f(P)\, d\mathbf{V} = \int_{\Gamma_1} f(P)\mathbf{V}_{N-1}(x - e_1)\, d\mathbf{v}_1.$$

Hence,

$$(d/dx) \int_{E<x} f(P)\, d\mathbf{V} = \int_{\Gamma_1} f(P)\Omega_{N-1}(x - e_1)\, d\mathbf{v}_1 = \int_{\Delta} \Omega_{N-1}(x - e_1)\, d\mathbf{v}_1,$$

and consequently,

$$\langle f \rangle = \mathsf{P}\{q_1, p_1 \in \Delta\} = \int_{\Delta} [\Omega_{N-1}(x - e_1)/\Omega_N(x)]\, d\mathbf{v}_1.$$

This means that if x is the total energy of the system, the particle we selected is distributed in its phase space Γ_1 with density

$$\varphi(q_1, p_1) = \Omega_{N-1}(x - e_1)/\Omega_N(x), \tag{6}$$

where $e_1 = e(q_1, p_1)$ is the energy of the particle at the given point. In particular, we see that this distribution depends only on the energy $e(q_1, p_1)$ of the particle, so that for all points of the space Γ_1 with the same value of the energy, the probability density will be the same. It is easy to show that (5) follows from (6) in a straightforward way.

§3. Distribution of the energy of a particle. Gibbs' theorem

First, we consider the problem of the energy distribution of an individual particle. As in §2 we define the probability $\mathsf{P}(e_1 < u)$ that the energy of the first particle is less than the positive number u, to be the microcanonical average of $f(P)$, where $f(P)$ is equal to 1 if, at the point P of the surface

$e_1 + \cdots + e_N = E$, we have $e_1 = e(q_1, p_1) < u$, and is equal to zero otherwise. Since $f(P)$ depends only on e_1, we can put

$$f(P) = \varphi_u(e_1) = \begin{cases} 1 & (e_1 < u) \\ 0 & (e_1 \geq u) \end{cases}$$

and, according to (5), we have

$$\begin{aligned} \mathsf{P}(e_1 < u) &= \langle\varphi_u(e_1)\rangle = [\Omega_N(x)]^{-1} \int_0^\infty \varphi_u(z)\omega(z)\Omega_{N-1}(E - z)\, dz \\ &= \int_0^u [\omega(z)\Omega_{N-1}(E - z)/\Omega_N(E)]\, dz, \end{aligned} \tag{7}$$

where E denotes the total energy of the system. We see that for an asymptotic estimate of the quantity $\mathsf{P}(e_1 < u)$ we must find approximate expressions for the quantities $\Omega_N(E)$ and $\Omega_{N-1}(E - z)$.

For this purpose let us use the method described in my book [1]. We assume that the number of particles N and the total energy E of the system are infinitely large, maintaining the constant ratio $E/N = a$. If we assume further that $\omega(z)$ satisfies the usual conditions ([1; p. 76]), then a unique positive number α exists for which

$$\int_0^\infty z\omega(z)e^{-\alpha z}\, dz \Big/ \int_0^\infty \omega(z)e^{-\alpha z}\, dz = a. \tag{8}$$

Let us put

$$\int_0^\infty \omega(z)e^{-\alpha z}\, dz = \lambda, \qquad \omega(z)e^{-\alpha z} = \lambda\varphi(z).$$

The function $\varphi(z)$ is the density of a certain distribution whose mathematical expectation, according to (8), is equal to a. Since $\omega(z) = \lambda e^{\alpha z}\varphi(z)$, relation (2) gives

$$\Omega_N(x) = \lambda^N e^{\alpha x}\Phi_N(x),$$

where

$$\Phi_N(x) = \int^{(N-1)} \left[\prod_{i=1}^{N-1} \{\varphi(z_i)\, dz_i\}\right] \varphi\left(x - \sum_{i=1}^{N-1} z_i\right). \tag{9}$$

Therefore, relation (7) can be rewritten in the form

$$\begin{aligned} \mathsf{P}(e_1 < u) &= \int_0^u [\lambda e^{\alpha z}\varphi(z)\lambda^{N-1}e^{\alpha(E-z)}\Phi_{N-1}(E - z)/\lambda^N e^{\alpha E}\Phi_N(E)]\, dz \\ &= \int_0^u [\varphi(z)\Phi_{N-1}(E - z)/\Phi_N(E)]\, dz. \end{aligned} \tag{10}$$

But $\Phi_N(x)$, by its definition (9), is the density at the point x of the distribution of the sum of N mutually independent terms with density $\varphi(z)$. According to the local limit theorem ([8; p. 228]), all of whose requirements are satisfied in our case, we have

$$(11)\quad \Phi_N[Na + z(Nb)^{\frac{1}{2}}] = \Phi_N[E + z(Nb)^{\frac{1}{2}}] = (2\pi Nb)^{-\frac{1}{2}} e^{-z^2/2Nb} + N^{-\frac{1}{2}}\eta(z),$$

where b denotes the dispersion of the distribution $\varphi(z)$, and where as $N \to \infty$, $\eta(z) \to 0$ uniformly on the entire real line $(-\infty < z < \infty)$. Thus,

$$\Phi_N(E) = (2\pi Nb)^{-\frac{1}{2}} + o(N^{-\frac{1}{2}}),$$

and for $z = O(1)$

$$\Phi_{N-1}(E - z) = (2\pi Nb)^{-\frac{1}{2}} + o(N^{-\frac{1}{2}}).$$

Therefore, for constant u

$$(12)\qquad \Phi_{N-1}(E - z)/\Phi_N(E) = 1 + o(1)$$

uniformly in the interval $0 \leq z \leq u$. From (10) we obtain as $N \to \infty$

$$\mathsf{P}(e_1 < u) = \int_0^u \varphi(z)\, dz + o(1),$$

and consequently,

$$\mathsf{P}(e_1 < u) \to \int_0^u \varphi(z)\, dz = \int_0^u \omega(z)e^{-\alpha z}\, dz \Big/ \int_0^\infty \omega(z)e^{-\alpha z}\, dz \qquad (N \to \infty).$$

Thus, the problem of determining the limit distribution of the energy of a particle is completely solved. Let us recall further that the number α is the unique root of equation (8). In the sequel we put

$$\int_0^u \omega(z)e^{-\alpha z}\, dz \Big/ \int_0^\infty \omega(z)e^{-\alpha z}\, dz = K(u),$$

so that for any $u \geq 0$

$$\mathsf{P}(e_1 < u) \to K(u) \qquad (N \to \infty).$$

The corresponding *local* limit theorem is also an immediate consequence of our calculations. Differentiating (10) and using the estimate (12) we find

$$d\mathsf{P}(e_1 < u)/du = \varphi(u)\Phi_{N-1}(E - u)/\Phi_N(E) = \varphi(u) + o(1).$$

Whence, as $N \to \infty$

$$d\mathsf{P}(e_1 < u)/du \to \varphi(u) = K'(u) \qquad (0 \leq u < \infty).$$

Further, according to (11), we obviously have $\Phi_{N-1}(x) = O(N^{-\frac{1}{2}})$ uniformly for $-\infty < x < \infty$. According to the asymptotic estimate we have obtained for $\Phi_N(E)$, there exists a constant $c > 0$ such that

$$\Phi_{N-1}(E - z)/\Phi_N(E) < c$$

uniformly with respect to all values of N and z. Since (10) shows that

$$\mathsf{P}(e_1 \geq u) = \int_u^\infty [\varphi(z)\Phi_{N-1}(E - z)/\Phi_N(E)]\, dz,$$

we find for all N and for all $u \geq 0$

$$\mathsf{P}(e_1 \geq u) < c \int_u^\infty \varphi(z)\, dz = c[1 - K(u)]. \tag{13}$$

Let us assume now that the function $\omega(z)e^{-\gamma z}$ is decreasing (at least for sufficiently large z) for any $\gamma > 0$. (This is always the case in real physical problems.) Then, for sufficiently large u and with $0 < \beta < \alpha$

$$1 - K(u) = \lambda^{-1} \int_u^\infty e^{-\alpha z}\omega(z)\, dz = \lambda^{-1} \int_u^\infty e^{-\beta z}\omega(z)e^{-(\alpha-\beta)z}\, dz$$

$$< \lambda^{-1}\omega(u)e^{-(\alpha-\beta)u} \int_u^\infty e^{-\beta z}\, dz = (\beta\lambda)^{-1}\omega(u)e^{-\alpha u} < e^{-\beta u},$$

and, consequently, for any N and sufficiently large u

$$\mathsf{P}(e_1 \geq u) < e^{-\beta u}.$$

Thus we arrive at the following general estimate which we will need later:

LEMMA. *Let $0 < \beta < \alpha$. Then for any N and sufficiently large u*

$$\mathsf{P}(e_1 \geq u) < e^{-\beta u}.$$

Formula (6), §2, for the density of the distribution of a particle in its phase space can also be approximated simply and conveniently with the help of these asymptotic formulas. First, we easily find

$$\Omega_{N-1}(E - e_1)/\Omega_N(E) = \lambda^{-1}e^{-\alpha e_1}\Phi_{N-1}(E - e_1)/\Phi_N(E).$$

Since $\Phi_{N-1}(E - e_1)/\Phi_N(E)$ tends to unity for constant e_1 and $N \to \infty$, then as $N \to \infty$,

$$\lim\, [\Omega_{N-1}(E - e_1)/\Omega_N(E)] = \lambda^{-1}e^{-\alpha e_1}.$$

Formula (6) shows therefore that when N and E approach infinity while maintaining a constant ratio, the density of the distribution of a particle in its phase space also tends [at the point (q_1, p_1) of this space] to the quantity

$$\lambda^{-1}e^{-\alpha e_1} = e^{-\alpha e_1} \Big/ \int_0^\infty \omega(z)e^{-\alpha z}\, dz,$$

where $e_1 = e(q_1, p_1)$ is the energy of the particle at the point (q_1, p_1). The probability that, for example, the first particle will belong to a certain domain Δ of its phase space will thus tend to the quantity

$$\int_\Delta e^{-\alpha e(q_1, p_1)}\, d\mathbf{v}_1 \Big/ \int_{\Gamma_1} e^{-\alpha e(q_1, p_1)}\, d\mathbf{v}_1,$$

where $d\mathbf{v}_1$ is the volume element of the space Γ_1. This result, which is usually called *Gibbs' Theorem*, can be interpreted as follows: If a large system is microcanonically distributed in its phase space, a small part of this system (a particle) is *canonically* distributed in its phase space.

§4. Derivation of the fundamental formula

We turn now to the solution of a more difficult problem. Let an arbitrary $u \geq 0$ be given. Then, in general, for an arbitrary point P of the surface

$$\textstyle\sum_{i=1}^N e(q_i, p_i) = E,$$

some particles will have energy less than u, and the others will have energy greater than or equal to u.

Let us denote by $\psi_N(u)$ the relative number of those particles whose energies are less than u, so that $\psi_N(u)$ is one of the numbers

$$k/N \qquad (k = 0, 1, \cdots, N).$$

Obviously, $\psi_N(u)$, for given N and u, is a function of the point P of the constant energy surface and consequently, from our point of view, must be considered as a random variable. We shall now find the distribution of this random variable. The solution of this problem is the key to the study of symmetric functions of the particle energies. In fact, at two points for which the values of $\psi_N(u)$ coincide for all $u \geq 0$, the energies of the particles obviously can differ from one another only in so far as the particles are enumerated in a different order. Conversely, this means that any symmetric function of the particle energies assumes identical values at two such points. In other words, any symmetric function of the particle energies can be considered as a functional whose argument is the function $\psi_N(u)$.

Such a functional (if it is continuous in a certain definite sense) will be "almost constant" on a surface of constant energy, if the function $\psi_N(u)$ possesses this property. (This point is discussed in greater detail in §9.) It is for precisely this reason that we must study the distribution law of the quantity $\psi_N(u)$.

The event $\psi_N(u) = k/N$ $(0 \leq k \leq N)$, whose probability we set out to determine, is the event in which k of the N particles of the system have energy less than u, while the remainder have energy greater than or equal to u.

Since k particles can be chosen in $C(N, k)$ different ways, and since all the particles are indistinguishable from one another,

$$\begin{aligned} &\mathsf{P}\{\psi_N(u) = k/N\} \\ (14) \qquad &= C(N, k)\,\mathsf{P}\{e_i < u\ (1 \le i \le k);\ e_i \ge u\ (k+1 \le i \le N)\} \\ &= C(N, k)\,\mathsf{P}\{A_k(u)\}, \end{aligned}$$

where, for brevity, we have used the symbol $A_k(u)$ to denote the event $\{e_i < u\ (1 \le i \le k);\ e_i \ge u\ (k+1 \le i \le N)\}$. If we now put, as in §3,

$$\varphi_u(z) = \begin{cases} 1 & (0 \le z < u), \\ 0 & (z < 0,\ z \ge u), \end{cases}$$

then the probability of the event $A_k(u)$ is by definition the microcanonical average of the quantity

$$F_k(u) = \left[\prod_{i=1}^{k} \varphi_u(e_i)\right] \prod_{i=k+1}^{N} [1 - \varphi_u(e_i)].$$

Therefore, if the total energy of the system is E, the application of (4), §2 yields

$$\begin{aligned} &\mathsf{P}\{A_k(u)\} = \langle F_k(u) \rangle = [\Omega_N(E)]^{-1} \int^{(N-1)} \left[\prod_{i=1}^{k} \varphi_u(z_i)\omega(z_i)\, dz_i\right] \\ (15) \qquad &\cdot \left[\prod_{i=k+1}^{N-1} [1 - \varphi_u(z_i)]\omega(z_i)\, dz_i\right]\left[1 - \varphi_u\left(E - \sum_{i=1}^{N-1} z_i\right)\right]\omega\left(E - \sum_{i=1}^{N-1} z_i\right). \end{aligned}$$

This gives us a very inconvenient expression for the probability

$$\mathsf{P}\{\psi_N(u) = k/N\}.$$

Hence, we shall make a preliminary transformation completely analogous to (although somewhat more complicated than) the one carried out in §3. Let us put

$$\lambda = \int_0^\infty \omega(z) e^{-\alpha z}\, dz,$$

where α is determined from the equation

$$\int_0^\infty z\omega(z) e^{-\alpha z}\, dz \Big/ \int_0^\infty \omega(z) e^{-\alpha z}\, dz = E/N = a.$$

Let us also put

$$K(u) = \lambda^{-1} \int_0^u \omega(z) e^{-\alpha z}\, dz,$$

$$\varphi_u(z)\omega(z)e^{-\alpha z} \Big/ \int_0^\infty \varphi_u(z)\omega(z)e^{-\alpha z}\, dz = [\lambda K(u)]^{-1} \varphi_u(z)\omega(z)e^{-\alpha z} = f_1(z),$$

and

$$[1 - \varphi_u(z)]\omega(z)e^{-\alpha z} \Big/ \int_0^\infty [1 - \varphi_u(z)]\omega(z)e^{-\alpha z}\, dz$$
$$= \lambda^{-1}[1 - K(u)]^{-1}[1 - \varphi_u(z)]\omega(z)e^{-\alpha z} = f_2(z),$$

so that $f_1(z)$ and $f_2(z)$ are the densities of certain distributions. Thus, we have

$$\varphi_u(z)\omega(z) = \lambda K(u) e^{\alpha z} f_1(z),$$
$$[1 - \varphi_u(z)]\omega(z) = \lambda[1 - K(u)] e^{\alpha z} f_2(z),$$

and (15) takes the form

$$\text{(16)} \quad \begin{aligned} \mathsf{P}\{A_k(u)\} &= \lambda^N [\Omega_N(E)]^{-1} [K(u)]^k [1 - K(u)]^{N-k} e^{\alpha E} \\ &\cdot \int^{(N-1)} \left[\prod_{i=1}^{k} f_1(z_i)\, dz_i\right] \left[\prod_{i=k+1}^{N-1} f_2(z_i)\, dz_i\right] f_2\left(E - \sum_{i=1}^{N-1} z_i\right). \end{aligned}$$

Here the $(N - 1)$-fold integral on the right side obviously represents the density, at the point E, of the distribution of the sum of N mutually independent random variables, of which the first k are distributed with density $f_1(z)$ and the remaining $N - k$ are distributed with density $f_2(z)$. For brevity, let us denote this density by $\Pi_k(u, E)$. Then (14) and (16) give

$$\text{(17)} \quad \begin{aligned} &\mathsf{P}\{\psi_N(u) = k/N\} \\ &\quad = C(N, k)\lambda^N e^{\alpha E} [\Omega_N(E)]^{-1} [K(u)]^k [1 - K(u)]^{N-k} \Pi_k(u, E). \end{aligned}$$

Finally, let us note that $\Pi_0(0, E)$ represents the density at the point E of the distribution of the sum of N mutually independent random variables distributed with density $\lambda^{-1}\omega(z)e^{-\alpha z}$. In §3 we denoted this quantity by $\Phi_N(E)$ and we saw that

$$\Omega_N(E) = \lambda^N e^{\alpha E} \Phi_N(E) = \lambda^N e^{\alpha E} \Pi_0(0, E).$$

Therefore, we find

$$\text{(18)} \quad \begin{aligned} &\mathsf{P}\{\psi_N(u) = k/N\} \\ &\quad = C(N, k)[K(u)]^k [1 - K(u)]^{N-k} \Pi_k(u, E)/\Pi_0(0, E). \end{aligned}$$

Formula (18) is basic to all our further calculations. We shall see that it is exceptionally convenient for this purpose. The study of (18) involves finding the asymptotic estimate of the density $\Pi_k(u, E)$. This computation naturally involves the distributions $f_1(z)$ and $f_2(z)$. Therefore we shall describe here some of the simplest properties of these distributions which will be needed later.

Let $M_1^{(r)}$ and $M_2^{(r)}$ denote, respectively, the rth moments of the distributions $f_1(z)$ and $f_2(z)$. Then

$$K(u)M_1^{(r)} = K(u) \int_0^\infty z^r f_1(z)\, dz = \lambda^{-1} \int_0^\infty \varphi_u(z) z^r \omega(z) e^{-\alpha z}\, dz$$

$$= \lambda^{-1} \int_0^u z^r \omega(z) e^{-\alpha z}\, dz,$$

and analogously,

$$[1 - K(u)]M_2^{(r)} = \lambda^{-1} \int_u^\infty z^r \omega(z) e^{-\alpha z}\, dz.$$

Whence,

$$K(u)M_1^{(r)} + [1 - K(u)]M_2^{(r)} = \lambda^{-1} \int_0^\infty z^r \omega(z) e^{-\alpha z}\, dz.$$

Thus, a linear combination of the rth moments of the distributions f_1 and f_2 with coefficients $K(u)$ and $1 - K(u)$ is always equal (independently of u) to the moment of the same order of the distribution

$$\lambda^{-1} \omega(z) e^{-\alpha z} = \varphi(z)$$

which we considered in §3. By the definition of the number α, the mathematical expectation of this law is equal to $a = E/N$. Its dispersion will be denoted by b. Thus,

$$K(u)M_1^{(1)} + [1 - K(u)]M_2^{(1)} = a, \tag{19}$$

and

$$K(u)M_1^{(2)} + [1 - K(u)]M_2^{(2)} = b + a^2. \tag{20}$$

§5. Distribution of the maximum and minimum energy of a particle

Before we proceed to the general asymptotic analysis of the basic formula (18), let us consider some of its more elementary applications. We shall show that with the help of this formula it is easy to find a very accurate expression for the distribution of the maximum and minimum energy of a particle (when the total energy of the system is fixed). [Here, and in the sequel, by maximum (minimum) energy of a particle is meant that energy which is the maximum (minimum) possessed by any single particle for a given state of the entire system. Since the division of the total energy among the separate particles depends on the state of the system, so does the maximum (minimum) energy of a particle.]

The requirement that the maximum energy of a particle be less than u is equivalent to the requirement that all N particles have energy less than u. This implies that $\psi_N(u) = 1$. Thus the distribution of the maximum energy of a particle is

$$G(u) = \mathsf{P}\{\psi_N(u) = 1\},$$

which can be found from (18) by putting $k = N$. Hence,

$$G(u) = \{K(u)\}^N \Pi_N(u, E)/\Pi_0(0, E). \tag{21}$$

We turn now to a detailed calculation whose result will express with great accuracy the limiting form of the distribution $G(u)$:

THEOREM. *For constant* x $(-\infty < x < \infty)$, *as* $N \to \infty$,

$$G(\alpha^{-1} \ln N\{1 + \ln \omega(\alpha^{-1} \ln N)/\ln N + x/\ln N\}) \to \exp(-\alpha^{-1}\lambda^{-1}e^{-x}).$$

In particular, this theorem shows that for sufficiently large N the maximum energy of a particle, with a probability as near as we like to unity, will be approximately

$$\alpha^{-1} \ln N + \alpha^{-1} \ln \omega(\alpha^{-1} \ln N).$$

Proof. For brevity let us put

$$\alpha^{-1} \ln N\{1 + \ln \omega(\alpha^{-1} \ln N)/\ln N + x/\ln N\} = u_N .$$

According to (21), the theorem will be proved if we show that, as $N \to \infty$,

$$\{K(u_N)\}^N \to \exp(-\alpha^{-1}\lambda^{-1}e^{-x}) \tag{*}$$

and

$$\Pi_N(u_N , E)/\Pi_0(0, E) \to 1. \tag{**}$$

As in §3, we shall suppose here that the function $\omega(z)e^{-\gamma z}$ decreases for any $\gamma > 0$ and for sufficiently large z. Moreover we shall make the further assumption that as $z \to \infty$

$$\omega'(z)/\omega(z) = o(z^{-1}). \tag{22}$$

[Both assumptions are always fulfilled in real physical problems where $\omega(z)$ is usually approximated by some power of the variable z. It is easy, however, to show that our first assumption is a consequence of the second.]

In order to prove relation (*), let us note that integration by parts gives

$$1 - K(u) = \lambda^{-1} \int_u^\infty \omega(z)e^{-\alpha z}\, dz$$

$$= (\lambda\alpha)^{-1}\omega(u)e^{-\alpha u} + (\lambda\alpha)^{-1} \int_u^\infty \omega'(z)e^{-\alpha z}\, dz.$$

But it follows from (22) that, as $z \to \infty$, $\omega'(z) = o[\omega(z)]$ and therefore as $u \to \infty$

$$\int_u^\infty \omega'(z)e^{-\alpha z}\, dz = o\left\{\int_u^\infty \omega(z)e^{-\alpha z}\, dz\right\} = o\{1 - K(u)\},$$

and we find

$$1 - K(u) = (\lambda\alpha)^{-1}\omega(u)e^{-\alpha u} + o\{1 - K(u)\}.$$

Hence, as $u \to \infty$

(23) $$1 - K(u) = (\lambda\alpha)^{-1}\omega(u)e^{-\alpha u}\{1 + o(1)\}.$$

Further, it follows from our first assumption that for any $\delta > 0$ and for sufficiently large z we have $\omega(z) < e^{\delta z}$ and, hence, $\ln \omega(z) < \delta z$. In other words we have $\ln \omega(y) = o(y)$ and consequently, in particular,

$$\ln \omega(\alpha^{-1} \ln N) = o(\ln N).$$

Whence,

$$1 + \ln \omega(\alpha^{-1} \ln N)/\ln N + x/\ln N = 1 + o(1) \quad (N \to \infty),$$

and thus,

$$u_N = \alpha^{-1} \ln N + o(\ln N).$$

Consequently,

$$\omega(u_N) = \omega\{\alpha^{-1} \ln N + o(\ln N)\}$$

$$= \omega(\alpha^{-1} \ln N) + o[\ln N\omega'(\alpha^{-1} \ln N)].$$

Therefore, by (22) as $N \to \infty$

(24) $$\omega(u_N) = \omega(\alpha^{-1} \ln N) + o[\omega(\alpha^{-1} \ln N)] \sim \omega(\alpha^{-1} \ln N).$$

On the other hand, we have

(25) $$e^{-\alpha u_N} = \exp[- \ln N - \ln \omega(\alpha^{-1} \ln N) - x] = e^{-x}[N\omega(\alpha^{-1} \ln N)]^{-1}.$$

From (23), (24) and (25) it follows that as $N \to \infty$

$$1 - K(u_N) \sim e^{-x}[\lambda\alpha N]^{-1},$$

$$(26) \qquad K(u_n) = 1 - e^{-x}[\lambda\alpha N]^{-1} + o(N^{-1}),$$

$$\{K(u_N)\}^N \to \exp(-e^{-x}/\lambda\alpha),$$

which proves relation (*).

Turning now to the relation

$$\Pi_N(u_N, E)/\Pi_0(0, E) \to 1,$$

we notice first of all that the denominator of this ratio, by definition, coincides with the quantity $\Phi_N(E)$, which we investigated in §3. We saw there that as $N \to \infty$

$$\Pi_0(0, E) = \Phi_N(E) \sim (2\pi N b)^{-\frac{1}{2}},$$

where b is the dispersion of the distribution whose density is

$$\varphi(z) = \lambda^{-1}\omega(z)e^{-\alpha z}.$$

Thus we have to investigate only the numerator $\Pi_N(u_N, E)$. This quantity is the density, at point E, of the distribution of the sum of N mutually independent random variables, each of which is distributed with the density (see §4)

$$f_1(z) = \varphi_{u_N}(z)\omega(z)e^{-\alpha z}[\lambda K(u_N)]^{-1}.$$

In §4 the mathematical expectation of this term was denoted by $M_1^{(1)}$. It equals

$$(27) \qquad M_1^{(1)} = [\lambda K(u_N)]^{-1}\int_0^{u_N} z\omega(z)e^{-\alpha z}\,dz,$$

and as $N \to \infty$ it approaches

$$(28) \qquad \lambda^{-1}\int_0^{\infty} z\omega(z)e^{-\alpha z}\,dz = a = E/N.$$

Therefore, the mathematical expectation of the sum will be near $Na = E$, i.e., just that point at which we must estimate the density of the distribution of this sum. In order to make this estimate we must establish the order of smallness of the difference $M_1^{(1)} - a$. According to (27) and (28), we have

$$\lambda[a - M_1^{(1)}] = \int_0^{\infty} z\omega(z)e^{-\alpha z}\,dz - [K(u_N)]^{-1}\int_0^{u_N} z\omega(z)e^{-\alpha z}\,dz$$

$$= \int_{u_N}^{\infty} z\omega(z)e^{-\alpha z}\,dz - [1 - K(u_N)][K(u_N)]^{-1}\int_0^{u_N} z\omega(z)e^{-\alpha z}\,dz.$$

By (26), the second term, as $N \to \infty$, has the form $O(N^{-1})$. As for the first term, integration by parts gives

$$\int_{u_N}^{\infty} z\omega(z)e^{-\alpha z}\,dz = \alpha^{-1}u_N\omega(u_N)e^{-\alpha u_N} + \alpha^{-1}\int_{u_N}^{\infty}[\omega(z) + z\omega'(z)]e^{-\alpha z}\,dz.$$

It is obvious from (23), (26) and $u_N = O(\ln N)$, that the first term of this formula has the form $O(N^{-1}\ln N)$. Using (22), we see that the second term is of order $O[1 - K(u_N)] = O(N^{-1})$. Thus,

$$a - M_1^{(1)} = O(N^{-1}\ln N),$$

and it follows that

$$E - NM_1^{(1)} = N[a - M_1^{(1)}] = O(\ln N). \tag{29}$$

If, for brevity, the dispersion $M_1^{(2)} - [M_1^{(1)}]^2$ of the law $f_1(z)$ is denoted by b_N, then the local limit theorem [see (11), §3] yields

$$\Pi_N(u_N, E) \sim (2\pi N b_N)^{-\frac{1}{2}}\exp(-[E - NM_1^{(1)}]^2/2Nb_N) \quad (N \to \infty).$$

It is obvious that, as $N \to \infty$, the dispersion b_N of the law $f_1(z)$ tends to the dispersion b of the law $\varphi(z)$. Since, by (29)

$$[E - NM_1^{(1)}]^2/2Nb_N = O(N^{-1}\ln^2 N),$$

we have

$$\Pi_N(u_N, E) \sim (2\pi N b_N)^{-\frac{1}{2}} \qquad (N \to \infty),$$

i.e., Π_N coincides asymptotically with $\Pi_0(0, E)$. This proves relation (**) and simultaneously completes the proof of our theorem.

Now we turn to the distribution of the minimum energy of a particle. This problem is solved analogously to the previous problem, but the calculation here is much simpler. First of all, it is obvious that the probability that the minimum energy of a particle is less than u is equal to

$$g(u) = 1 - \mathsf{P}\{\psi_N(u) = 0\}.$$

In order to get an asymptotic estimate of $g(u)$, we can again use our basic formula (18), this time putting $k = 0$,

$$\mathsf{P}\{\psi_N(u) = 0\} = \{1 - K(u)\}^N \Pi_0(u, E)/\Pi_0(0, E).$$

Let us denote by $\mathbf{v}(u)$ the volume of that part of the phase space of a particle in which its energy is less than u. Then, by the definition of the function $\omega(z)$,

$$\mathbf{v}(u) = \int_0^u \omega(z)\,dz.$$

The function $\mathbf{v}(u)$ is continuous and increasing. Let us denote by $u(\mathbf{v})$ the inverse function which, consequently, is also continuous and increasing for any $\mathbf{v} > 0$. Then the following limit theorem solves our problem.

THEOREM. *For any constant* $x > 0$, *and as* $N \to \infty$,

$$g\{u(x/N)\} \to 1 - \exp(-x/\lambda).$$

Proof. Obviously, the theorem will be proved if we can show that, as $N \to \infty$

$$(*) \qquad [1 - K\{u(x/N)\}]^N \to \exp(-x/\lambda)$$

and

$$(**) \qquad \Pi_0\{u(x/N), E\}/\Pi_0(0, E) \to 1.$$

We note that, as $u \to 0$

$$K(u) = \lambda^{-1} \int_0^u \omega(z) e^{-\alpha z}\, dz = \lambda^{-1} \int_0^u \omega(z)[1 + O(z)]\, dz$$
$$= \lambda^{-1}\mathbf{v}(u)[1 + O(u)].$$

In particular, as $N \to \infty$

$$K\{u(x/N)\} = \lambda^{-1}(x/N)[1 + o(1)].$$

From this we easily establish relation (*).

In order to prove relation (**), we note first of all that the denominator $\Pi_0(0, E)$ is exactly the same as in the preceding theorem. We know that as $N \to \infty$ it is equivalent to $(2\pi Nb)^{-\frac{1}{2}}$. Thus, again, we have to estimate only the numerator $\Pi_0\{u(x/N), E\}$. We shall not carry out the estimate explicitly as the calculation is very similar to that of the preceding theorem. The only difference is that this calculation is much easier. In this case the basic problem is to estimate the difference $M_2^{(1)} - a$. We easily find that as $N \to \infty$

$$M_2^{(1)} - a = O(N^{-1}).$$

The application of the local limit theorem now shows, exactly as in the preceding theorem, that as $N \to \infty$

$$\Pi_0\{u(x/N), E\} \sim (2\pi Nb)^{-\frac{1}{2}}.$$

This proves relation (**) and hence the entire theorem.

§6. The basic limit theorem

Let us turn now to the general asymptotic analysis of (18), §4, which will be used to establish the most important limit theorem of the theory.

First we make a few preliminary remarks. The quantity $N\psi_N(u)$ denotes the number of particles whose energies are less than u. Formula (18) gives the probability that this number is equal to k. According to the results of §3, for large N the probability that the energy of an individual particle is less than u is near $K(u)$. If this probability were exactly $K(u)$ and *if the energies of the different particles, as random variables, were mutually independent*, then we would be dealing with a simple case of Bernoulli trials: $N\psi_N(u)$ would simply be the number of successes in N independent trials of a certain event $(e_i < u)$ whose probability for each individual trial is $K(u)$. The probability that this number of successes equal k, would be

$$C(N, k)\{K(u)\}^k\{1 - K(u)\}^{N-k},$$

which is just the first factor of the right side of (18). However, the particle energies are dependent since their sum must equal the total energy $E = Na$. The second factor

$$\Pi_k(u, E)/\Pi_0(0, E)$$

of the right side of (18) is a "correction factor" which takes account of the mutual dependence of the energies of the particles. Since this dependence is very weak (we have a large number N of random variables e_i related by the one equation $\sum_i e_i = Na$), we can expect that the correction factor in many cases will be nearly unity. For example, in both problems of §5 we saw that it tends to unity under appropriate conditions. (This is precisely why the problems of §5 were solvable by relatively elementary methods.)

We now perform the detailed calculation. In this section we consider the number u to be constant and therefore, for brevity, we write K instead of $K(u)$. Otherwise we retain the notation of the preceding section.

The mathematical expectation of the quantity $\psi_N(u)$ is obviously equal to $\mathsf{P}(e_i < u)$, which (see §3) is nearly equal to K. We shall consider at first only those values of k for which k/N is nearly equal to this mathematical expectation and thus nearly equal to K. Let us put

$$(k/N) - K = h.$$

We shall assume in the following that as $N \to \infty$

$$h = O(N^{-\frac{1}{2}}).$$

By using the (local) Laplace-de Moivre theorem, we find for the first factors of the right side of (18) the asymptotic expression

$$\begin{aligned} &C(N, k)K^k(1 - K)^{N-k} \\ &\quad = [1 + o(1)][2\pi NK(1 - K)]^{-\frac{1}{2}} \exp\,[-Nh^2/2K(1 - K)]. \end{aligned} \tag{30}$$

Consider now the last (correction) factor. Its denominator, which is independent of k and u, is equal to

$$\Pi_0(0, E) = [1 + o(1)](2\pi Nb)^{-\frac{1}{2}}, \tag{31}$$

where b denotes the dispersion of the distribution whose density is

$$\varphi(z) = \lambda^{-1}\omega(z)e^{-\alpha z}.$$

Thus, for the asymptotic analysis of (18) for values of k/N near K, we have to investigate only the behavior of the numerator $\Pi_k(u, E)$ of the correction factor. Let us recall that $\Pi_k(u, E)$ is the density, at the point $E = Na$, of the distribution of the sum s_N of N mutually independent random variables, of which the first k are distributed according to the law $f_1(z)$, and the remaining $N - k$ are distributed according to the law $f_2(z)$ (see §4). The mathematical expectations of these two laws were denoted in §4 by $M_1^{(1)}$ and $M_2^{(1)}$. For brevity we shall denote them here by M_1 and M_2.

The second moments of these distributions will be denoted by $M_1^{(2)}$ and $M_2^{(2)}$, and their dispersions by D_1 and D_2, so that

$$D_1 = M_1^{(2)} - M_1^2, \qquad D_2 = M_2^{(2)} - M_2^2.$$

Denoting the mathematical expectation and the dispersion of the sum s_N by $\langle s_N \rangle$ and Ds_N, respectively, and using formulas (19) and (20) of §4, we have

$$\begin{aligned}\langle s_N \rangle &= kM_1 + (N - k)M_2 = N[(K + h)M_1 + (1 - K - h)M_2] \\ &= N[KM_1 + (1 - K)M_2 + h(M_1 - M_2)] = N(a + h\Delta),\end{aligned}$$

where $\Delta = M_1 - M_2$. Further, for $h = O(N^{-\frac{1}{2}})$

$$\begin{aligned}Ds_N &= kD_1 + (N - k)D_2 \\ &= N[(K + h)(M_1^{(2)} - M_1^2) + (1 - K - h)(M_2^{(2)} - M_2^2)] \\ &= N[KM_1^{(2)} + (1 - K)M_2^{(2)} - KM_1^2 - (1 - K)M_2^2 + O(N^{-\frac{1}{2}})] \\ &= N[b + a^2 - KM_1^2 - (1 - K)M_2^2 + O(N^{-\frac{1}{2}})] = ND_0 + O(N^{\frac{1}{2}}),\end{aligned}$$

where

$$D_0 = b + a^2 - KM_1^2 - (1 - K)M_2^2.$$

Hence, D_0 is constant for constant u.

To get an estimate of the quantity $\Pi_k(u, E)$ we can now apply the local limit theorem since all of the conditions are satisfied (see [8; p. 228]). We find that

$$\text{(32)}\qquad \begin{aligned}\Pi_k(u, E) &= [1 + o(1)](2\pi D s_N)^{-\frac{1}{2}} \exp\left(-[E - \langle s_N \rangle]^2/2Ds_N\right)\\ &= [1 + o(1)](2\pi N D_0)^{-\frac{1}{2}} \exp\left(-Nh^2\Delta^2/2D_0\right).\end{aligned}$$

Using the estimates (30), (31) and (32), we can put (18) in the form

$$\mathsf{P}\{\psi_N(u) = k/N\}$$
$$= [1 + o(1)]b^{\frac{1}{2}}[2\pi NK(1 - K)D_0]^{-\frac{1}{2}} \exp\left[-Nh^2\{1/2K(1 - K) + \Delta^2/2D_0\}\right],$$

where $h = (k/N) - K = O(N^{-\frac{1}{2}})$. Let us put

$$K(1 - K)D_0/b = D.$$

Then, under the same assumption, we obtain

$$\mathsf{P}\{\psi_N(u) = k/N\}$$
$$= [1 + o(1)](2\pi ND)^{-\frac{1}{2}} \exp\left[-Nh^2\{D_0 + K(1 - K)\Delta^2\}/2bD\right].$$

Finally, it follows from (19) of §4 that

$$\begin{aligned}D_0 + K(1 - K)\Delta^2 &= b + a^2 - KM_1^2 - (1 - K)M_2^2\\ &\qquad + K(1 - K)(M_1^2 + M_2^2 - 2M_1M_2)\\ &= b + a^2 - [KM_1 + (1 - K)M_2]^2\\ &= b.\end{aligned}$$

Consequently, the dispersion of $\psi_N(u)$ is DN^{-1}, and we have

$$\text{(33)}\qquad \begin{aligned}\mathsf{P}\{\psi_N(u) &= k/N\}\\ &= [1 + o(1)](2\pi ND)^{-\frac{1}{2}} \exp\left(-Nh^2/2D\right),\\ &\qquad [h = (k/N) - K = O(N^{-\frac{1}{2}})].\end{aligned}$$

This relation gives us the desired local limit theorem. Since the estimate given by (33) obviously holds uniformly for $A/N^{\frac{1}{2}} < h < B/N^{\frac{1}{2}}$, where A and B ($A < B$) are any constants, we can use the classical method to find the corresponding integral form:

THEOREM. *Let A and B ($A < B$) be any constants. Then for constant $u > 0$ and $N \to \infty$*

$$\mathsf{P}\{AN^{-\frac{1}{2}} < \psi_N(u) - K(u) < BN^{-\frac{1}{2}}\} \to \pi^{-\frac{1}{2}} \int_{A/(2D)^{\frac{1}{2}}}^{B/(2D)^{\frac{1}{2}}} e^{-z^2}\, dz,$$

where D depends on u, but not on N.

Using simple transformations we easily obtain the following more convenient form for D:

$$\text{(34)}\qquad D = K(1 - K)\{1 - \Delta^2 K(1 - K)/b\}.$$

We can infer from the beginning of this section that if the energies of the particles could be considered mutually independent, then the dispersion of $\psi_N(u)$ would asymptotically equal $K(1 - K)N^{-1}$. Formula (34) shows, therefore, that taking account of the mutual dependence of the energies of the various particles always *decreases* the dispersion of $\psi_N(u)$. This could have been foreseen since the relation $\sum_i e_i = E$, which relates the energies of the particles, obviously leads to a negative mutual correlation between the energies of any pair of particles.

As in the case of the classical Laplace-de Moivre theorem, it follows immediately from the proof of our theorem that the limit relation holds uniformly with respect to A and B $(-\infty < A < B < \infty)$. In particular, as $N \to \infty$ we have

$$(35)\qquad \begin{aligned} &\mathsf{P}\{\psi_N(u) - K(u) > AN^{-\frac{1}{2}}\} \to \pi^{-\frac{1}{2}} \int_{A/(2D)^{\frac{1}{2}}}^{\infty} e^{-z^2}\, dz, \\ &\mathsf{P}\{\psi_N(u) - K(u) < BN^{-\frac{1}{2}}\} \to \pi^{-\frac{1}{2}} \int_{-\infty}^{B/(2D)^{\frac{1}{2}}} e^{-z^2}\, dz. \end{aligned}$$

Finally, as the Bernoulli theorem (the law of large numbers) follows directly from the Laplace-de Moivre theorem, so our theorem obviously has the following corollary.

COROLLARY. *For any* $\epsilon > 0$, *no matter how small, and for any constant* u *and as* $N \to \infty$

$$(36)\qquad \mathsf{P}\{\,|\psi_N(u) - K(u)| > \epsilon\} \to 0.$$

§7. Probability that the energy of a particle lies in a given interval

Previously the particle energies were denoted by $e_1, \cdots, e_N$ and were arranged in an arbitrary order. Now we shall suppose that these energies are arranged in order of increasing magnitude, so that, for example, e_1 stands for the minimum, and e_N for the maximum energy of a particle.

In §5 we studied the distributions of e_1 and e_N. Using the basic limit theorem proved in §6 we can study, to the same accuracy, the probabilities that particle energies lie in given intervals. We shall devote the present section to this study.

Let us denote by u_k the positive number defined by the equation

$$K(u_k) = k/N$$

and let us put

$$u_k + AN^{-\frac{1}{2}} = u', \qquad u_k + BN^{-\frac{1}{2}} = u'',$$

where A and B $(A < B)$ are arbitrary constants. Then, obviously

$$
\begin{aligned}
\mathsf{P}\{u' < e_k < u''\} &= \mathsf{P}\{\psi_N(u') < k/N \le \psi_N(u'')\} \\
&= 1 - \mathsf{P}\{\psi_N(u') \ge k/N\} - \mathsf{P}\{\psi_N(u'') < k/N\} \qquad (37)\\
&= 1 - \mathsf{P}\{\psi_N(u') \ge K(u_k)\} - \mathsf{P}\{\psi_N(u'') < K(u_k)\}.
\end{aligned}
$$

In order to estimate the probabilities which appear on the right side of this equation, let us note that

$$
\begin{aligned}
K(u') &= K(u_k) + (u' - u_k)K'(u_k) + O\{(u' - u_k)^2\} \\
&= K(u_k) + AN^{-\frac{1}{2}}K'(u_k) + O(N^{-1}).
\end{aligned}
$$

Whence,

$$K(u_k) = K(u') - AN^{-\frac{1}{2}}K'(u_k) + O(N^{-1}).$$

Analogously,

$$K(u_k) = K(u'') - BN^{-\frac{1}{2}}K'(u_k) + O(N^{-1}).$$

Therefore, we have

$$\mathsf{P}\{\psi_N(u') \ge K(u_k)\} = \mathsf{P}\{\psi_N(u') - K(u') \ge -N^{-\frac{1}{2}}AK'(u_k) + O(N^{-1})\}$$

$$\mathsf{P}\{\psi_N(u'') < K(u_k)\} = \mathsf{P}\{\psi_N(u'') - K(u'') < -N^{-\frac{1}{2}}BK'(u_k) + O(N^{-1})\}$$

According to the limit relations (35), §6, we easily find, as $N \to \infty$

$$\mathsf{P}\{\psi_N(u') \ge K(u_k)\} - \pi^{-\frac{1}{2}} \int_{-AK'(u_k)(2D)^{-\frac{1}{2}}}^{\infty} e^{-z^2}\,dz \to 0,$$

$$\mathsf{P}\{\psi_N(u'') < K(u_k)\} - \pi^{-\frac{1}{2}} \int_{-\infty}^{-BK'(u_k)(2D)^{-\frac{1}{2}}} e^{-z^2}\,dz \to 0,$$

where D depends, of course, on u_k (and thus on k). The character of this dependence was established in §6. Comparing these relations with (37), we easily find

$$\mathsf{P}\{u_k + AN^{-\frac{1}{2}} < e_k < u_k + BN^{-\frac{1}{2}}\} - \pi^{-\frac{1}{2}} \int_{AK'(u_k)(2D)^{-\frac{1}{2}}}^{BK'(u_k)(2D)^{-\frac{1}{2}}} e^{-z^2}\,dz \to 0,$$

$$(N \to \infty).$$

We see that for large values of N, the energy e_k is approximately distributed according to a normal law with mean u_k and dispersion

$$D(u_k)/N[K'(u_k)]^2.$$

More definite conclusions can, of course, be obtained only by making some special assumptions about the growth of the index k as $N \to \infty$. The simplest and most natural assumption is that the ratio k/N tends to a certain definite limit λ $(0 < \lambda < 1)$, as $N \to \infty$. Then, determining the number u_0 from the equation

$$K(u_0) = \lambda,$$

we can assert that as $N \to \infty$, the distribution of the quantity $N^{\frac{1}{2}}(e_k - u_0)$ tends to a normal law with mean 0 and with constant dispersion

$$D(u_0)/[K'(u_0)]^2.$$

§8. A functional limit theorem

In §6 we established that for any constant value of u $(0 < u < \infty)$, the random variable $\psi_N(u)$ tends in probability to $K(u)$ as $N \to \infty$. This is just the content of the corollary of the basic limit theorem we introduced at the end of §6. Now our problem is to establish the fact that *the function $\psi_N(u)$ on its entire range (i.e., for $0 < u < \infty$), for large N and with very high probability, is arbitrarily close to the function $K(u)$.* In order to impart to this assertion an accurate meaning, it is of course necessary to define accurately the mutual distance of the functions $\psi_N(u)$ and $K(u)$. Since both functions are distributions on the half-line $(0, \infty)$, it is natural to look for a convenient general definition of distance between two such distributions. Let $F_1(u)$ and $F_2(u)$ be two distributions on the half-line $(0, \infty)$. The simplest definition of the distance between them is the upper bound

$$\sup_{0<u<\infty} | F_1(u) - F_2(u) |,$$

and it would be easy to show that this distance between the functions $\psi_N(u)$ and $K(u)$ actually tends to 0 in probability as $N \to \infty$. However, these two functions are actually much closer; and since this fact is important for our calculations, it will be more advantageous to introduce a somewhat different definition of the distance between two distributions. Let us put

$$\rho_\beta(F_1, F_2) = \sup_{0<u<\infty}\{ | F_1(u) - F_2(u) | e^{\beta u}\},$$

where β is a positive constant. This definition will give us, of course, a certain one-parameter family of distances. Two functions, for which $\rho_\beta(F_1, F_2)$ is small, must be very near to one another for large values of u since for $\rho_\beta(F_1, F_2) < \epsilon$ we have

$$| F_1(u) - F_2(u) | < \epsilon e^{-\beta u}.$$

We can now prove that if β is sufficiently small, then for large N the distance $\rho_\beta(\psi_N, K)$ becomes arbitrarily small with a very high probability, and consequently the functions $\psi_N(u)$ and $K(u)$ become extremely close to one another. In particular, we can prove the following proposition:

THEOREM. *For any* $\epsilon > 0$ *and* β $(0 \leq \beta < \alpha)$,

$$\mathsf{P}\{\rho_\beta(\psi_N, K) > \epsilon\} \to 0 \qquad (N \to \infty).$$

First we must prove the following lemma.

LEMMA. *Let* $\epsilon > 0$ *and* $\beta < \alpha$. *Then as* $u \to \infty$, *we have*

$$\mathsf{P}\{\sup_{v>u} [(1 - \psi_N(v))e^{\beta v}] > \epsilon\} \to 0$$

uniformly with respect to N.

Proof of the lemma. Let $\beta < \gamma < \alpha$. Using the lemma of §3, we have

$$\mathsf{P}\{e_1 \geq u\} < e^{-\gamma u}$$

for any N and for sufficiently large u. Therefore, the series (the order of particle enumeration is again arbitrary)

$$\sum_{k=0}^{\infty} e^{\beta k}\mathsf{P}(e_1 \geq k)$$

converges. Since

$$\mathsf{P}(e_1 \geq u) = 1 - \mathsf{E}\{\psi_N(u)\},$$

the series

$$\sum_{k=0}^{\infty} e^{\beta k}[1 - \mathsf{E}\{\psi_N(k)\}]$$

also converges.

Let $\epsilon e^{-\alpha} = \epsilon'$ and, for $k = 0, 1, 2, \cdots$, let

$$\mathsf{P}\{1 - \psi_N(k) > \epsilon' e^{-\beta k}\} = p_k .$$

Since the Chebyshev inequality yields

$$p_k \leq (1/\epsilon')e^{\beta k}[1 - \mathsf{E}\{\psi_N(k)\}],$$

the series $\sum_{k=0}^{\infty} p_k$ also converges. From this we conclude that

$$\mathsf{P}\{\sup_{k\geq r} [(1 - \psi_N(k))e^{\beta k}] > \epsilon'\} \leq \sum_{k=r}^{\infty} p_k \to 0 \quad (r \to \infty). \tag{38}$$

If for any $k \geq 0$

$$[1 - \psi_N(k)]e^{\beta k} < \epsilon',$$

then for $k < u < k + 1$

$$\psi_N(u) \geq \psi_N(k) > 1 - \epsilon' e^{-\beta k},$$

and, thus,

$$1 - \psi_N(u) < \epsilon' e^{-\beta k} < \epsilon' e^{-\beta(u-1)} < \epsilon e^{-\beta u}.$$

Hence it follows from (38) that

$$\mathsf{P}\{\sup_{v>r}[(1-\psi_N(v))e^{\beta v}] > \epsilon\} \to 0 \qquad (r \to \infty),$$

uniformly with respect to N. This proves the lemma.

Proof of the theorem. Let $A > 0$ be so large that

$$\mathsf{P}\{\sup_{v>A}[(1-\psi_N(v))e^{\beta v}] > \epsilon\} < \epsilon/2 \tag{39}$$

for any N. This is possible in view of the above lemma. In proving the lemma of §3, we saw that for $\beta < \gamma < \alpha$ and for sufficiently large v

$$1 - K(v) < e^{-\gamma v} < \epsilon e^{-\beta v}.$$

Therefore, for sufficiently large v it follows from

$$1 - \psi_N(v) \leq \epsilon e^{-\beta v}$$

that

$$|\psi_N(v) - K(v)| < \epsilon e^{-\beta v}.$$

From (39) it further follows that for sufficiently large A

$$\mathsf{P}\{\sup_{v>A}[|\psi_N(v) - K(v)|e^{\beta v}] > \epsilon\} < \epsilon/2. \tag{40}$$

Now let s be an integer greater than $2e^{\beta A}/\epsilon$. For any integer r $(0 \leq r \leq s)$ let us define the positive number u_r by the equation

$$K(u_r) = (r/s)K(A),$$

so that

$$0 = u_0 < u_1 < \cdots < u_s = A.$$

According to the corollary (36) of the basic limit theorem of §6, we have for sufficiently large N

$$\mathsf{P}\{|\psi_N(u_r) - K(u_r)| \geq (\epsilon/2)e^{-\beta A}\} < \epsilon/2s \qquad (r = 1, \cdots, s),$$

and, consequently [considering the fact that $\psi_N(0) = K(0) = 0$],

$$\mathsf{P}\{\sup_{0\leq r\leq s}|\psi_N(u_r) - K(u_r)| \geq (\epsilon/2)e^{-\beta A}\} < \epsilon/2. \tag{41}$$

However, if $0 \leq r < s$, and if both inequalities

$$|\psi_N(u_r) - K(u_r)| < (\epsilon/2)e^{-\beta A},$$

$$|\psi_N(u_{r+1}) - K(u_{r+1})| < (\epsilon/2)e^{-\beta A}$$

hold, then for $u_r \leq u \leq u_{r+1}$,

$$\begin{aligned} K(u) - K(A)/s - (\epsilon/2)e^{-\beta A} &\leq K(u_{r+1}) - K(A)/s - (\epsilon/2)e^{-\beta A} \\ &= K(u_r) - (\epsilon/2)e^{-\beta A} \leq \psi_N(u_r) \leq \psi_N(u) \\ &\leq \psi_N(u_{r+1}) < K(u_{r+1}) + (\epsilon/2)e^{-\beta A} \\ &= K(u_r) + K(A)/s + (\epsilon/2)e^{-\beta A} \\ &\leq K(u) + K(A)/s + (\epsilon/2)e^{-\beta A}. \end{aligned}$$

Consequently,

$$| \psi_N(u) - K(u) | < (\epsilon/2)e^{-\beta A} + 1/s < \epsilon e^{-\beta A} \leq \epsilon e^{-\beta u}.$$

Therefore, it follows from (41) that

$$\mathsf{P}\{\sup_{v \leq A} [| \psi_N(v) - K(v) | e^{\beta v}] > \epsilon\} < \epsilon/2.$$

Comparing this inequality with inequality (40), we see that for sufficiently large N

$$\mathsf{P}\{\rho_\beta(\psi_N, K) > \epsilon\} = \mathsf{P}\{\sup_{0<v<\infty}[| \psi_N(v) - K(v) | e^{\beta v}] > \epsilon\} < \epsilon.$$

This proves our theorem. In this way we show that for sufficiently large N the function $\psi_N(u)$ (the assignment of which is equivalent to the assignment of the whole set of particle energies $e_1, \cdots, e_N$ without regard to order) is over its entire range extremely close to the function $K(u)$ for the overwhelming majority of points of the given constant energy surface. The form of $K(u)$ depends only on the structure of the particles composing the system (and in particular does not depend on N).

§9. Continuous symmetric functions

Now we turn to the basic problem of our study—the behavior of symmetric functions of the particle energies on the constant energy surfaces of the system. We have in mind those functions $f(e_1, \cdots, e_N)$ of the particle energies which do not depend on the order of the $e_1, \cdots, e_N$, i.e., those functions which are unchanged by any permutation of the arguments.

It is obvious that the value of such a function at any point on the constant energy surface $e_1 + \cdots + e_N = E$ is completely determined by the value, at this point, of the function $\psi_N(u)$, whose properties we developed in the last section. Thus, we can consider a symmetric function $f(e_1, \cdots, e_N)$ as a *functional* whose argument is the distribution of $\psi_N(u)$. Such a point of view is especially convenient for us. In fact our goal is the study of the behavior of $f(e_1, \cdots, e_N)$ as $N \to \infty$ (under the condition that $e_1 + \cdots + e_N = Na$). Thus, we are actually dealing not with one function f, but with a whole family of such functions (one for each N). It is therefore first necessary to define accurately the meaning of $f(e_1, \cdots, e_N)$ for any N. It is simplest, of course, to do this by considering f as a functional with argument $\psi_N(u)$, and defining this functional directly for any discrete distribution $\psi_N(u)$ with any finite number N of possible values, if all these values have the same probability $1/N$ and if the arithmetical average of these values is equal to the constant a. Actually, as we shall see, it will be convenient (and to a certain degree necessary) to give to such a functional a much broader definition which includes the set of all distributions on the half-line $(0, \infty)$. So, for example, if

$$f(e_1, \cdots, e_N) = (1/N)(e_1^k + \cdots + e_N^k),$$

where k is a positive constant, we have

$$f(e_1, \cdots, e_N) = \int_0^\infty u^k \, d\psi_N(u),$$

and there is no reason why we cannot consider the functional

$$\int_0^\infty u^k \, d\psi(u) \tag{42}$$

to be defined for any distribution $\psi(u)$ for which this integral converges.

Of course, this whole method is possible only if $f(e_1, \cdots, e_N)$ is symmetric for any N.

Hence, we shall consider functionals $F[\psi]$ defined at least for all distributions of the form $\psi_N(u)$. We shall call such a functional (and also the corresponding symmetric function of the particle energies) *continuous*, if the difference $F[\varphi] - F[\psi]$ becomes infinitely small when the distance $\rho(\varphi, \psi)$ between the distributions φ and ψ tends to zero. It is of course clear that this definition of continuity of the functional depends critically on the way in which we define $\rho(\varphi, \psi)$. If we take, in particular, the definition

$$\rho_\beta(\varphi, \psi) = \sup_{0<u<\infty}[\,|\varphi(u) - \psi(u)|\, e^{\beta u}]$$

which we used in §8, then any functional $F[\psi]$, which is continuous for some value of the parameter β, will obviously be continuous for all greater values of β. It is easy to compute that for any $k > 0$ the linear functional (42) is continuous for $\beta > 0$ and discontinuous at $\beta = 0$. It is just this fact that the majority of the simplest and most useful functionals are discontinuous at $\beta = 0$ (i.e., when the distance between the distributions φ and ψ is defined simply as the upper bound of the function $|\varphi(u) - \psi(u)|$), which forces us to introduce the distance ρ_β. This definition is very convenient for the development of the theory and, moreover, broadens considerably the class of continuous functionals.

Our goal is to prove that all functionals $F[\psi]$, continuous for some (not very large) value of β, possess the property of "near-constancy" on constant energy surfaces. This means that at the overwhelming majority of points of the surface

$$\sum_{i=1}^{N} e(q_i, p_i) = Na,$$

we must show that the functional $F[\psi_N(u)]$ assumes values arbitrarily close (if N is sufficiently large) to a certain constant (which is independent of N and the choice of point).

Using the theorem of §8, we see that this number will be $F[K]$, i.e., the value assumed by $F[\psi]$ for $\psi = K(u)$. Of course, $F[\psi]$ must be defined and continuous at $\psi = K(u)$. Actually, it is an immediate consequence of the theorem of §8, and our definition of continuity, that the following theorem is true.

THEOREM. *Let the functional $F[\psi]$ be defined and continuous for $\psi = K(u)$ [and, of course, for all $\psi = \psi_N(u)$] for some $\beta < \alpha$. Then, for any $\epsilon > 0$, no matter how small, as $N \to \infty$*

$$\mathsf{P}\{ \mid F[\psi_N] - F[K] \mid > \epsilon\} \to 0.$$

This theorem establishes the "near-constancy" of continuous symmetric functions of the particle energies on multi-dimensional surfaces of constant energy.

§10. Generalizations of Gibbs' theorem

At the end of §3 we proved Gibbs' theorem. This states that the microcanonical distribution of a system on the constant energy surface

$$\sum_{i=1}^{N} e(q_i, p_i) = E = Na$$

(which we now denote by S_N) implies, in the limit $N \to \infty$, the so-called *canonical* distribution for an individual particle in its phase space Γ_1. This theorem is of fundamental importance since it gives a theoretical foundation to the widely used and very convenient canonical method of averaging for a system in contact with a heat bath (i.e., freely exchanging energy with practically infinitely large surroundings). This theorem provides the method of canonical averaging with a rigorous justification. Thus, canonical averaging is no longer just a convenient analytical device whose validity is established only empirically.

In this theory, the microcanonical distribution assumption for the "large" system remains as a hypothetical and incompletely justified feature. Using the results of the last section, we can, to a certain degree, free ourselves from this assumption. Instead of the microcanonical distribution hypothesis we can limit ourselves to a much more general assumption, which requires only that the distribution of the system be symmetric with respect to its constituent particles. (Of course, we must also assume that the distributions are, in a very broad sense, continuous and bounded.)

To establish this result, we retain all the notation of the preceding section. Let us say that the functional $F[\psi]$, which we considered in §9, is of *positive* type, if $F[\psi] \geq 0$ is always true and if $F[K] > 0$. We shall call $F[\psi]$ *bounded*, if there exists a number $Q > 0$, independent of N, such that the microcanonical average of the quantity $F^2[\psi_N]$ on the surface S_N satisfies

$$[\Omega_N(E)]^{-1} \int_{S_N} (F^2[\psi_N]/\text{grad } E)\, d\Sigma_N < Q, \tag{43}$$

where $d\Sigma_N$ is the "surface" element of the surface S_N, and $\Omega_N(E)$ is the "area" of the total surface. These properties are possessed by the functionals most frequently met in applications, and, in particular, by all functionals of the form (42).

The microcanonical distribution of the system on the surface S_N is completely characterized by noting that its density is proportional to $1/\text{grad } E$. We now consider a broad class of distributions which are characterized by densities proportional to the expression

$$F[\psi_N]/\text{grad } E,$$

where $F[\psi]$ is any positive bounded functional, continuous for some $\beta < \alpha$. (We have a microcanonical distribution if $F[\psi] \equiv 1$.)

Each distribution of the system on the surface S_N implies a certain completely defined distribution for each particle in its own phase space. For example, the distribution of the first particle can be described by giving the probability $\mathsf{P}(A_1 \in \Delta)$ that the point A_1 of the phase space Γ_1, representing the state of the first particle, belongs to some domain Δ of this space. When the system is microcanonically distributed on the surface S_N, Gibbs' theorem tells us that as $N \to \infty$

$$\mathsf{P}\{A_1 \in \Delta\} \to \int_\Delta e^{-\alpha e(q_1, p_1)}\, d\mathbf{v}_1 \Big/ \int_{\Gamma_1} e^{-\alpha e(q_1, p_1)}\, d\mathbf{v}_1,$$

(see the end of §3). We now prove that this limit relation holds for the much broader assumptions which we have made. In particular, we prove the following theorem:

THEOREM. *Let $F[\psi]$ be a positive bounded functional, continuous for some $\beta < \alpha$, and defined for all $\psi_N(u)$ and for $K(u)$. If the system is distributed on the surface S_N with a density proportional to the quantity*

$$F[\psi_N]/\text{grad } E,$$

then in the limit an individual particle will be distributed canonically, i.e., for any domain Δ of the space Γ_1, as $N \to \infty$

$$\mathsf{P}\{A_1 \in \Delta\} \to \int_\Delta e^{-\alpha e(q_1, p_1)}\, d\mathbf{v}_1 \Big/ \int_{\Gamma_1} e^{-\alpha e(q_1, p_1)}\, d\mathbf{v}_1 .$$

Proof. Let us first introduce certain definitions. Let $S_N(\Delta)$ denote the set of all points of the surface S_N for which $A_1 \in \Delta$ (i.e., the set of all Hamiltonian variables for which the state of the first particle defines a point A_1 of the space Γ_1 belonging to the domain Δ of Γ_1). The probability $\mathsf{P}(A_1 \in \Delta)$ will be denoted by $\mathsf{P}_m(A_1 \in \Delta)$ if the distribution of the system on the surface S_N is microcanonical, and by $\mathsf{P}_s(A_1 \in \Delta)$ if the distribution satisfies the conditions of our theorem. We obviously have

$$\begin{aligned} \mathsf{P}_m\{A_1 \in \Delta\} &= \left\{\int_{S_N(\Delta)} d\Sigma_N/\operatorname{grad} E\right\} \Big/ \left\{\int_{S_N} d\Sigma_N/\operatorname{grad} E\right\} \\ &= [\Omega_N(E)]^{-1} \int_{S_N(\Delta)} d\Sigma_N/\operatorname{grad} E, \end{aligned} \tag{44a}$$

$$\mathsf{P}_s\{A_1 \in \Delta\} = \left\{\int_{S_N(\Delta)} F[\psi_N]\, d\Sigma_N/\operatorname{grad} E\right\} \Big/ \left\{\int_{S_N} F[\psi_N]\, d\Sigma_N/\operatorname{grad} E\right\}. \tag{44b}$$

For brevity, let us put

$$\left\{\int_{S_N} F[\psi_N]\, d\Sigma_N/\operatorname{grad} E\right\}^{-1} = \lambda_N, \qquad F[K] = F^*,$$

so that $F^* > 0$ as in the hypotheses of the theorem.

We have, by (44b),

$$\begin{aligned} \mathsf{P}_s\{A_1 \in \Delta\} &= \lambda_N \int_{S_N(\Delta)} F[\psi_N]\, d\Sigma_N/\operatorname{grad} E \\ &= \lambda_N F^* \int_{S_N(\Delta)} d\Sigma_N/\operatorname{grad} E + \lambda_N \int_{S_N(\Delta)} (F - F^*)\, d\Sigma_N/\operatorname{grad} E. \end{aligned} \tag{45}$$

Let S_N' and S_N'' denote, respectively, the parts of the surface S_N at which $| F[\psi_N] - F^* | < \epsilon$ and $| F[\psi_N] - F^* | \geq \epsilon$ where ϵ is an arbitrary positive constant less than $F^*/2$. In an analogous way we define the parts $S_N'(\Delta)$ and $S_N''(\Delta)$ in the domain $S_N(\Delta)$. Let us also put

$$\int_{S_N'} d\Sigma_N/\operatorname{grad} E = \Omega_N'(E), \qquad \int_{S_N''} d\Sigma_N/\operatorname{grad} E = \Omega_N''(E).$$

Using the theorem of §9, we have

$$\mathsf{P}_m\{ | F[\psi_N] - F^* | > \epsilon\} = \Omega_N''(E)/\Omega_N(E) \to 0 \quad (N \to \infty). \tag{46}$$

Further, it is obvious that

$$\begin{aligned} &\lambda_N \left| \int_{S_N'(\Delta)} (F - F^*)\, d\Sigma_N/\operatorname{grad} E \right| \\ &\qquad < \epsilon\lambda_N \int_{S_N} d\Sigma_N/\operatorname{grad} E = \epsilon\lambda_N\Omega_N(E). \end{aligned} \tag{47}$$

On the other hand, since $\epsilon < F^*/2$,

$$\lambda_N^{-1} \geq \int_{S_N'} F[\psi_N]\, d\Sigma_N/\text{grad } E = F^*\Omega_N'(E) + \int_{S_N'} (F - F^*)\, d\Sigma_N/\text{grad } E$$
$$\geq F^*\Omega_N'(E) - \epsilon\Omega_N'(E) > \tfrac{1}{2}F^*\Omega_N'(E).$$

But by (46)

$$\Omega_N'(E)/\Omega_N(E) > \tfrac{1}{2},$$

for sufficiently large N. Therefore,

$$\lambda_N^{-1} > \tfrac{1}{4}F^*\Omega_N(E).$$

Whence,

$$\lambda_N\Omega_N(E) < 4/F^*. \tag{48}$$

In this case, it follows from (47) that

$$\left| \lambda_N \int_{S_N'(\Delta)} (F - F^*)\, d\Sigma_N/\text{grad } E \right| < 4\epsilon/F^*. \tag{49}$$

Further, for any Δ and sufficiently large N, (48) and (46) give

$$\begin{aligned} \lambda_N \int_{S_N''(\Delta)} F^*\, d\Sigma_N/\text{grad } E &\leq \lambda_N F^* \int_{S_N''} d\Sigma_N/\text{grad } E \\ &= \lambda_N F^*\Omega_N''(E) = \lambda_N F^*\Omega_N(E)[\Omega_N''(E)/\Omega_N(E)] < \epsilon. \end{aligned} \tag{50}$$

Finally,

$$\begin{aligned} \lambda_N \int_{S_N''(\Delta)} F[\psi_N]\, d\Sigma_N/\text{grad } E &\leq \lambda_N \int_{S_N''} F[\psi_N]\, d\Sigma_N/\text{grad } E \\ &= \lambda_N \int_{\substack{S_N'' \\ (F<1/\epsilon)}} F\, d\Sigma_N/\text{grad } E + \lambda_N \int_{\substack{S_N'' \\ (F\geq 1/\epsilon)}} F\, d\Sigma_N/\text{grad } E \\ &< (\lambda_N/\epsilon)\Omega_N''(E) + \lambda_N \int_{S_N''} \epsilon F^2\, d\Sigma_N/\text{grad } E \\ &\leq (\lambda_N/\epsilon)\Omega_N(E)[\Omega_N''(E)/\Omega_N(E)] + \epsilon\lambda_N \int_{S_N} F^2\, d\Sigma_N/\text{grad } E. \end{aligned}$$

If N is so large that

$$\Omega_N''(E)/\Omega_N(E) < \epsilon^2,$$

then by (48)

$$\lambda_N \int_{S_N''(\Delta)} F[\psi_N]\, d\Sigma_N/\text{grad } E < 4\epsilon/F^* + 4\epsilon Q/F^* = (Q + 1)4\epsilon/F^*. \tag{51}$$

We now collect our results. From (50) and (51) we have

$$\left| \lambda_N \int_{S_N''(\Delta)} (F - F^*)\, d\Sigma_N/\text{grad } E \right| < \epsilon\{4(Q+1)/F^* + 1\}. \tag{52}$$

But by (49) and (52), it follows from (45) that

$$\left| \mathsf{P}_s(A_1 \in \Delta) - \lambda_N F^* \int_{S_N(\Delta)} d\Sigma_N/\text{grad } E \right| < B\epsilon, \tag{53}$$

where B is a positive constant. This inequality is fulfilled for sufficiently large N no matter what the domain Δ of the space Γ_1. For $\Delta = \Gamma_1$,

$$\mathsf{P}_s(A_1 \in \Delta) = 1,$$

and we find

$$| 1 - \lambda_N F^* \Omega_N(E) | < B\epsilon.$$

Since ϵ is as small as we like,

$$\lambda_N \Omega_N(E) \to 1/F^* \qquad (N \to \infty),$$

or

$$\lambda_N F^* = (1 + \alpha_N)/\Omega_N(E), \qquad \alpha_N \to 0 \qquad (N \to \infty).$$

Turning to (53) and using (44a), we therefore find

$$| \mathsf{P}_s(A_1 \in \Delta) - (1 + \alpha_N)\mathsf{P}_m(A_1 \in \Delta) | < B\epsilon.$$

This means that as $N \to \infty$

$$\mathsf{P}_s(A_1 \in \Delta) - \mathsf{P}_m(A_1 \in \Delta) \to 0,$$

and consequently, according to Gibbs' theorem,

$$\mathsf{P}_s(A_1 \in \Delta) \to \int_\Delta e^{-\alpha e(q_1, p_1)}\, d\mathbf{v}_1 \bigg/ \int_{\Gamma_1} e^{-\alpha e(q_1, p_1)}\, d\mathbf{v}_1,$$

which was to be proved.

REFERENCES

1. Khinchin, A. Ia. *Mathematical Foundations of Statistical Mechanics*, New York, Dover, 1949.
2. Editor's Note: For further discussion of quantum-mechanical ergodic theorems see, for example, D. Ter Haar, Revs. Modern Phys. **27,** 289 (1955).
3. von Mises, R. *Wahrscheinlichkeitsrechnung und ihre Anwendung in der Statistik und theoretischen Physik*, Leipzig und Wien, 1931.
4. Feller, W. *An Introduction to Probability Theory and Its Applications*, 2nd ed., New York, Wiley, 1958.
5. Gnedenko, B. V. *On a local limit theorem of the theory of probability*, Uspekhi Mat. Nauk **3,** 187 (1949).
6. Meisler, D. G., Parasyuk, O. S., and Rvacheva, E. L. *On a multi-dimensional local limit theorem of the theory of probability*, Ukr. Mat. Zhurnal, no. 1 (1949).
7. For details of the elementary properties of point lattices see, for example, D. Hilbert and S. E. Cohn-Vossen, *Geometry and the Imagination*, New York, Chelsea, 1952.
8. The reader interested in problems of this type may refer to B. V. Gnedenko and A. N. Kolmogorov, *Limit Distributions for Sums of Independent Random Variables*, Cambridge, Addison-Wesley, 1954.
9. Khinchin, A. Ia. *On the analytical apparatus of physical statistics*, Trudy Mat. Inst. im. Steklov **33,** 3 (1950).
10. Trudy Mat. Inst. im. Steklov **38,** 345 (1951).
11. Fowler, R. H. *Statistical Mechanics*, Cambridge, 1929.
12. von Mises, R. *Les lois de probabilité pour les fonctions statistiques*, Ann. Inst. H. Poincaré **6,** 185 (1936).
13. von Mises, R. *On the asymptotic distribution of differentiable statistical functions*, Ann. Math. Statistics **18,** 309 (1947).
14. *Pamiati Alexander Alexandrovich Andranov*, Moscow, 1955.

INDEX

A CATALOG OF SELECTED
DOVER BOOKS
IN SCIENCE AND MATHEMATICS

A CATALOG OF SELECTED
DOVER BOOKS
IN SCIENCE AND MATHEMATICS

QUALITATIVE THEORY OF DIFFERENTIAL EQUATIONS, V.V. Nemytskii and V.V. Stepanov. Classic graduate-level text by two prominent Soviet mathematicians covers classical differential equations as well as topological dynamics and ergodic theory. Bibliographies. 523pp. 5⅜ x 8½. 65954-2 Pa. $14.95

MATRICES AND LINEAR ALGEBRA, Hans Schneider and George Phillip Barker. Basic textbook covers theory of matrices and its applications to systems of linear equations and related topics such as determinants, eigenvalues and differential equations. Numerous exercises. 432pp. 5⅜ x 8½. 66014-1 Pa. $10.95

QUANTUM THEORY, David Bohm. This advanced undergraduate-level text presents the quantum theory in terms of qualitative and imaginative concepts, followed by specific applications worked out in mathematical detail. Preface. Index. 655pp. 5⅜ x 8½. 65969-0 Pa. $14.95

ATOMIC PHYSICS (8th edition), Max Born. Nobel laureate's lucid treatment of kinetic theory of gases, elementary particles, nuclear atom, wave-corpuscles, atomic structure and spectral lines, much more. Over 40 appendices, bibliography. 495pp. 5⅜ x 8½. 65984-4 Pa. $13.95

ELECTRONIC STRUCTURE AND THE PROPERTIES OF SOLIDS: The Physics of the Chemical Bond, Walter A. Harrison. Innovative text offers basic understanding of the electronic structure of covalent and ionic solids, simple metals, transition metals and their compounds. Problems. 1980 edition. 532pp. 6⅛ x 9¼. 66021-4 Pa. $16.95

BOUNDARY VALUE PROBLEMS OF HEAT CONDUCTION, M. Necati Özisik. Systematic, comprehensive treatment of modern mathematical methods of solving problems in heat conduction and diffusion. Numerous examples and problems. Selected references. Appendices. 505pp. 5⅜ x 8½. 65990-9 Pa. $12.95

A SHORT HISTORY OF CHEMISTRY (3rd edition), J.R. Partington. Classic exposition explores origins of chemistry, alchemy, early medical chemistry, nature of atmosphere, theory of valency, laws and structure of atomic theory, much more. 428pp. 5⅜ x 8½. (Available in U.S. only) 65977-1 Pa. $11.95

A HISTORY OF ASTRONOMY, A. Pannekoek. Well-balanced, carefully reasoned study covers such topics as Ptolemaic theory, work of Copernicus, Kepler, Newton, Eddington's work on stars, much more. Illustrated. References. 521pp. 5⅜ x 8½. 65994-1 Pa. $12.95

PRINCIPLES OF METEOROLOGICAL ANALYSIS, Walter J. Saucier. Highly respected, abundantly illustrated classic reviews atmospheric variables, hydrostatics, static stability, various analyses (scalar, cross-section, isobaric, isentropic, more). For intermediate meteorology students. 454pp. 6⅛ x 9¼. 65979-8 Pa. $14.95

DE RE METALLICA, Georgius Agricola. The famous Hoover translation of greatest treatise on technological chemistry, engineering, geology, mining of early modern times (1556). All 289 original woodcuts. 638pp. 6¾ x 11. 60006-8 Pa. $21.95

SOME THEORY OF SAMPLING, William Edwards Deming. Analysis of the problems, theory and design of sampling techniques for social scientists, industrial managers and others who find statistics increasingly important in their work. 61 tables. 90 figures. xvii + 602pp. 5⅜ x 8½. 64684-X Pa. $16.95

THE VARIOUS AND INGENIOUS MACHINES OF AGOSTINO RAMELLI: A Classic Sixteenth-Century Illustrated Treatise on Technology, Agostino Ramelli. One of the most widely known and copied works on machinery in the 16th century. 194 detailed plates of water pumps, grain mills, cranes, more. 608pp. 9 x 12. 28180-9 Pa. $24.95

LINEAR PROGRAMMING AND ECONOMIC ANALYSIS, Robert Dorfman, Paul A. Samuelson and Robert M. Solow. First comprehensive treatment of linear programming in standard economic analysis. Game theory, modern welfare economics, Leontief input-output, more. 525pp. 5⅜ x 8½. 65491-5 Pa. $14.95

ELEMENTARY DECISION THEORY, Herman Chernoff and Lincoln E. Moses. Clear introduction to statistics and statistical theory covers data processing, probability and random variables, testing hypotheses, much more. Exercises. 364pp. 5⅜ x 8½. 65218-1 Pa. $10.95

THE COMPLEAT STRATEGYST: Being a Primer on the Theory of Games of Strategy, J.D. Williams. Highly entertaining classic describes, with many illustrated examples, how to select best strategies in conflict situations. Prefaces. Appendices. 268pp. 5⅜ x 8½. 25101-2 Pa. $7.95

CONSTRUCTIONS AND COMBINATORIAL PROBLEMS IN DESIGN OF EXPERIMENTS, Damaraju Raghavarao. In-depth reference work examines orthogonal Latin squares, incomplete block designs, tactical configuration, partial geometry, much more. Abundant explanations, examples. 416pp. 5⅜ x 8¼. 65685-3 Pa. $10.95

THE ABSOLUTE DIFFERENTIAL CALCULUS (CALCULUS OF TENSORS), Tullio Levi-Civita. Great 20th-century mathematician's classic work on material necessary for mathematical grasp of theory of relativity. 452pp. 5⅜ x 8½. 63401-9 Pa. $11.95

VECTOR AND TENSOR ANALYSIS WITH APPLICATIONS, A.I. Borisenko and I.E. Tarapov. Concise introduction. Worked-out problems, solutions, exercises. 257pp. 5⅜ x 8¼. 63833-2 Pa. $8.95

THE FOUR-COLOR PROBLEM: Assaults and Conquest, Thomas L. Saaty and Paul G. Kainen. Engrossing, comprehensive account of the century-old combinatorial topological problem, its history and solution. Bibliographies. Index. 110 figures. 228pp. 5⅜ x 8½. 65092-8 Pa. $7.95